WILD OTTERS
Predation and Populations

Wild Otters

Predation and Populations

HANS KRUUK

Institute of Terrestrial Ecology
Banchory, Scotland

Drawings by Diana Brown

Oxford   New York   Tokyo
OXFORD UNIVERSITY PRESS
1995

Oxford University Press, Walton Street, Oxford OX2 6DP

Oxford   New York
Athens   Auckland   Bangkok   Bombay
Calcutta   Cape Town   Dar es Salaam   Delhi
Florence   Hong Kong   Istanbul   Karachi
Kuala Lumpur   Madras   Madrid   Melbourne
Mexico City   Nairobi   Paris   Singapore
Taipei   Tokyo   Toronto
and associated companies in
Berlin   Ibadan

Oxford is a trade mark of Oxford University Press

Published in the United States
by Oxford University Press Inc., New York

A catalogue record for this book is available from the British Library

Library of Congress Cataloging in Publication Data
Kruuk, H. (Hans)
Wild otters: predation and populations/Hans Kruuk; drawings by Diana Brown.
Includes bibliographical references.
1. Lutra lutra—Scotland.   I. Title.
QL737.C25K795   1995   599.74'447—dc20   94–34842
ISBN   0 19 854070 1

Typeset by EXPO Holdings, Malaysia
Printed in Great Britain on acid-free paper by
St Edmundsbury Press, Bury St Edmunds

To
Loeske and Johnny

Preface

From the beginning of my interest in otters I have been determined to observe directly what these animals did in the wild. I wanted to watch them in their natural environment, to see them feeding, interacting, reproducing, dying, to get beyond the usual mammalogists' occupation of analyses of faeces and distribution of 'spraints'. I have been fortunate indeed that I could spend several years trying to achieve this, helped and advised by many people. Results were a long time coming: it was like building up an investment portfolio, which would only pay a dividend after several years. Otters are difficult animals, therefore the investment had to be high, and the pay-offs sporadic—but the results, when they did come, were fascinating.

When writing about a species like the otter, one cannot separate the interests of science and of conservation: the questions asked by the academic about the evolution of social organization need much the same kind of data as does the request for a species-management plan from a conservation organization. Any scientist interested in the ecology and behaviour of otters has to be a conservationist. With this understanding, this book is written mostly for the scientist–naturalist, with conservation issues never far away.

Inevitably, there were statistical problems which I had to take into account—but in the text the discussion of those has been kept to a minimum. Any reference to details of statistical significance can easily be skipped by the reader, and anyone who is specially interested is referred to the relevant papers in journals. I have tried to make this a book about the animals, not just figures.

Much of our field work on otter ecology has been done along the shores of Shetland, and this led to the later research in freshwater areas in north-east Scotland and elsewhere. These beautiful, rich, and wild places have many unique characteristics, some of which will be discussed here. But over and above the local problems, I hope that I have been able to focus sufficiently on the general relationships between otters and their environment to contribute to the understanding of otter ecology everywhere, not just in Shetland and in Scotland. Much remains to be studied in order to understand the huge variability of the behaviour of these animals in their natural habitat; this book is not the final word, but I hope it will make a contribution towards that understanding.

The studies reported here are the joint effort of a team, of frequently changing composition. Many friends, colleagues, and students have

helped, either by their observations, by their assistance in the field or in
the lab, with their advice or their comments and criticisms, or in some
other way. It is impossible to mention everybody, but I would like to
express special gratitude to some of them for the many services ren-
dered: David Carss, Jim Conroy, Peter Dale, Leon Durbin, Ulco
Glimmerveen, Addie de Jongh, Andrew Moorhouse, Bart Nolet, Erik-Jan
Ouwerkerk, Dennis Wansink, all of whom contributed much to the field
work. At the Institute in Banchory I had much help from Brian Staines,
Donnie Littlejohn, Phil Bacon, Geraldine McGowan, and John Morris,
and my daughter Loeske made useful suggestions on the manuscript.
Both Ray Hewson and especially David Carss read and very helpfully
commented on the entire completed manuscript, and Diana Brown's
fine and exciting illustrations go a long way to capture the spirit of
otters in this book. Thomas Stephan let me use some of his excellent
photographs (Figs 1.7 and 6.2). To all I am deeply grateful.

More than anybody else my wife Jane provided essential back-up
during field work, support, ideas, suggestions on and criticism of the
text, and much help with its preparation: she has made it all possible.

Banchory   H.K.
March 1994

Contents

1

Introduction

*Otters (*Lutra lutra *L.) are still common in some areas of Europe, but in many parts they have disappeared. I ask questions about what limits their numbers; to that end, otter densities and social organization are studied, the behaviour which maintains them, factors affecting mortality and reproduction, and the effects of food and habitat. More general principles, such as the effects of resource dispersion, the biological function of scent marking, and the energetics of hunting play an important part. Most of the research discussed in this book was carried out in Shetland, some in north-east Scotland, and the areas are described. Various field techniques for watching and studying otters are detailed. A brief background is given for* Lutra lutra, *its distribution and some previous studies, and the other species of otters (*Lutrinae) are mentioned, with their main regions of occurrence and some of the published research.*

BACKGROUND

Shetland 1988, a balmy April morning. The Sound in front of me was like a sheet of blue silk, with just the distant murmur of moving water as a reminder that there were tides here. Miles away across the huge expanse of sea lay Yell, with its gently curving outline of low yellow hills. I could hear the crooning of a raft of eider ducks, and a gannet hurried past: a scene close to perfection.

An otter swam out, a small dark head at the sharp end of a V in the water. A dive, the tail flipping vertically, then an ever-increasing circle with a few bubbles gently spluttering up in the centre. I knew it would come up again in about the same place, perhaps in 15 or 20 seconds, perhaps earlier if it had caught a fish. There was something totally predictable about the animal; it belonged there, beautifully shaped as it was for this expanse of water, and one of many in Shetland.

Yet, at about the same time that I was sitting there along the Yell Sound, the last otter disappeared from Holland, the land where I grew up with its sea, lakes, dykes, and rivers and, as I knew it, a paradise for otters if ever there was one. A handful of otters remain in England, and they are still fairly common in Spain, Portugal, Greece, northern Scandinavia, Eastern Europe, and some other places. But in hundreds of miles elsewhere in Europe, where the animals were abundant a few decades ago, now there are none.

For this disaster, which has hit otters more than most other mammals in Europe, blame is often put on the doorstep of that catch-all evil, pollution. However, the scientific evidence is only circumstantial, and many other factors could also play a crucial role. Other species affected by pollutants in the 1950s and 1960s, the heron, sparrow hawk, peregrine, have now returned to their old haunts, mink are thriving along many banks; otters, however, are still faring badly.

What is needed to get to the basis of the problem is a clear answer to the ecological question which is essential to the conservation of every species: what limits numbers? Why are there now no more otters along the banks of a clean river in England or Holland, or for that matter, why are there not more otters here in Shetland than the few I see around me? These are questions which can be asked for otters anywhere, but they are relevant especially there where the problems bite, where the species is slipping away.

However, to answer them for otters, we need to know much more about the animals than we do. Firstly and essentially in the earlier years of my study, therefore, it was necessary to collect base-line information, in areas where the animals were apparently safe and plentiful. Now, with knowledge gained in Shetland and in the north-east of Scotland, we are in a better position to proceed with questions about otters in the rivers and marshes of mainland Britain and elsewhere. This is what this book is about.

The urgent conservation problem is one reason for my interest in otters. There are others too; I make no excuse for being infatuated with the animals, as is almost everyone else who has watched them.

There is no doubt that they are enormously appealing to a large public, even to many people who have seen them only on their television screens. Apart from all that, though, otters have an important scientific attraction, especially when juxtaposed with other species of carnivore.

In this book I want first to describe the way in which otters are organized, their social system and their numbers, and the methods we used to assess these. To explain what we found, to answer the all-important question about otter numbers, I need then to look at the resources used by otters, at the peculiar need of sea-foraging otters for freshwater, how they use their holts, and to examine in detail their relationship with prey species. With this information I try to interpret otters' population characteristics, their mortality and reproduction, and the important role of pollutants. Often, insights are gained by direct comparisons between sea-living and freshwater otters, focusing on relations with prey but also including observations on habitat and pollution problems. Many of these observations have important implications for the way in which we should manage the environment, in order to secure otter populations where there still are any, or wherever the animals were once common and are now gone.

Otters show themselves as a curiously vulnerable species, with a life history so relatively insecure that in some ways it is surprising not that they have gone from many of their haunts, but that they are still present at all along many banks and coasts.

QUESTIONS OF BEHAVIOUR AND ECOLOGY OF OTTERS

The problems which a conservationist would raise about a species such as the otter relate to numbers—how many, where, which habitats—and reasons for their decline. As a biologist, I want to know about the causes underlying the distribution of otters, and about their foraging mechanisms and the complicated interactions with their prey, and about the effects of the environment on their population dynamics. There is no conflict between the demands of conservation and science, but there is a strong incentive to phrase our questions in such a way that they address both.

The chief objectives, therefore, could be summarized by the following:

1. To describe the organization and densities of otters, and the behavioural mechanisms through which they are maintained, as well as factors affecting mortality and reproduction,

2. To discover the effects of food on numbers and organization, by studying diet, foraging behaviour, and strategies, the costs and benefits of foraging, and assessing prey populations,

3. To study the effects of habitat on numbers and dispersion.

The reasons for most research on otters are questions asked about the otter *per se*, as a species, or about otters in a particular area or population. But there are broader, biological principles underlying many of these problems, such as the relationship between territoriality and resource utilization, the relationship between animal numbers and availability of their food, the biological function of scent marking, the role of parental behaviour in the development of hunting skills, the energetics of hunting, the role of pollution as a cause of mortality, and many more. It is important that the contributions which these various otter studies make to our knowledge of these principles escape from the clutches of 'speciesism', and that they are seen as a further endowment of our knowledge of biology in general.

The central problem addressed here, the relationship between resources and animal numbers, has been a focus in ecology for many years, critically addressed at an early stage by David Lack in his reviews, in his own studies on birds (Lack 1954), and since then developed in great detail. An important aspect of the mechanism of this relationship is spatial organization, the pattern of dispersion of a species in relation to the environment. Especially in carnivores, in predatory species exploiting prey with complicated antipredator systems and organizations, these spatial aspects are crucial for the understanding of the impacts of resources on the predators.

Mammals, and especially carnivores, are highly variable in the way in which they distribute themselves in their habitat, with large differences between species, but also strong variation within species. To my knowledge there has been only one major hypothesis which tries to explain this variation in ecological terms, addressing environmental variation which correlates with differences in social organization. This is the Resource Dispersion Hypothesis (RDH), for carnivores in general, first put forward by Macdonald (1983) and refined later (Kruuk and Macdonald 1985; Carr and Macdonald 1986; Bacon *et al.* 1991). The RDH states that the sizes of home ranges and the number of members of the same species, conspecifics, inhabiting each range are determined not just by the overall density of food in the area, but by the dispersion, the patchiness, of the prey. Range size would be determined by the distances between individual patches, the numbers of animals inhabiting a range determined by the productivity of patches. Evidence for this comes from studies of foxes (Macdonald 1981), badgers (Kruuk 1978), hyaenas (Kruuk 1972), and others, almost all species which have group territories, that is where several adults share the same range.

There are some problems with the RDH, not the least of which is that there is as yet no good alternative. The hypothesis has not yet been tested with predictions, and no attempt has been made, for instance, to predict what spatial organization would be optimal for a given dispersion of prey. Also, there are phenomena in the spatial organization even of species which were central to the formulation of RDH, which cannot be explained by it (Latour 1988; Kruuk 1989a). One difficulty is that we know relatively little about RDH as it applies to solitary species, since there are remarkably few studies of solitary carnivores focusing on their social organization, territory size, and resources (reviewed by Sandell (1989)). The same is true for solitary species in other mammalian orders, such as ungulates (reviews in Jarman (1974)) and Clutton-Brock (1982)) and primates (Dunbar 1988). One of the aims of this study of otters, therefore, was to test some of the predictions of the RDH in a solitary carnivore.

Given otters' semi-aquatic way of life and all the special adaptations to this (Estes 1989), there are several interesting consequences which feature in the relationship between the animal and its resources. These include the fact that the home range or territory of otters has an almost linear shape along the water's edge (Erlinge 1967; Melquist and Hornocker 1983; Green et al. 1984) There is also the need for animals to dive into a distinctly hostile environment to get food (where cold threatens and oxygen is short) (Kramer 1988), so foraging is energetically very demanding. Otters have to cope with the extreme agility and mobility of their prey species, and there are several peculiar features in otter population dynamics. As I will demonstrate, all these phenomena have important bearings on resource utilization, and they are relevant when we test the RDH.

The question of what limits the numbers of otters in any one area can be approached from two different angles: from one direction, we can first study the effects of factors such as food by estimating availability, and quantifying foraging effort, and furthermore determine habitat utilization and possible restrictive mechanisms therein. From another angle, one can look at otter populations, at patterns and causes of mortality (including factors such as pollution and food shortage), and changes in reproductive success, and deduce possible mechanisms in population dynamics from age structures. Here I have attempted to use both approaches, and to bring these together by asking questions about individual behaviour as well as about populations.

STUDY AREAS

This book would probably never have been written if it had not been for my first few years in the wonderful study area in Shetland, at Lunna, and I will describe some of its details to provide a background for many of the results mentioned here. At the time, about 1983, too little was known about otters to tackle them confidently in places where they were scarce. With a 'difficult' species, few in numbers, persecuted and shy, usually nocturnal, intelligent, one needs a good deal of experience to be able to watch it, catch it, and to ask the right questions. This experience is hard to come by in a low-density area, where all the problems related to shyness are exacerbated. Shetland provided a superb answer to such predicaments: otters are common, they live along open, easily accessible, treeless coasts and perhaps most important of all, they are active during the day.

One could object to the use of a study area for otters which is so different from most other places where the animals occur: it is 'atypical', not representative, one cannot extrapolate results to elsewhere. The animals swim in seawater, in daytime, they have only one bank instead of two, the vegetation is different, as well as many other factors. Obviously, I have to be careful in using the Shetland results elsewhere, but that is true for every other individual study area as well. The point is that in Shetland we could devise hypotheses for the many questions which we had to ask about the otters, and these hypotheses could be tested elsewhere as well as in Shetland itself. The comparisons between the coastal ecosystems of Shetland and those in rivers and lakes were helpful, and indeed the fact alone that in freshwater areas we could use the practical experience gained in those Northern Isles fully justified our first choice of study area.

The general ecology of the islands and of the otter habitats in Shetland have been described by Watson (1978), Berry and Johnston (1980), Herfst (1984), Kruuk *et al.* (1989), and Kruuk and Moorhouse (1991). Shetland is a group of more than 100 islands at the northernmost tip of Britain, around 60° N—which is about the same latitude as the southern tip of Greenland. The total area is some 1400 km², of which about 980 km² is on the island of Mainland. These figures seem largely irrelevant, however, because of the very irregular coastlines, with deep inlets everywhere. Nowhere in Shetland is further than 5 km from the sea (Fig. 1.1).

The landscape of Shetland has a strange fascination for any field naturalist, with its rich, wild beauty and virtual independence from man's interference. It is rugged, barren, treeless, and gale-ridden, one

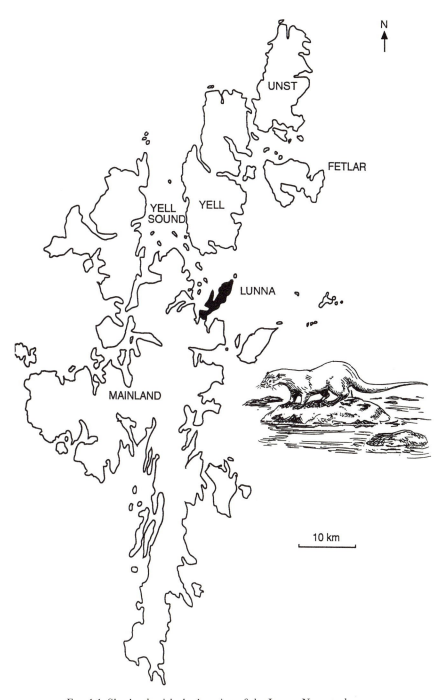

Fig. 1.1 Shetland, with the location of the Lunna Ness study area.

of the windiest and dampest places in Britain. It is totally different from Orkney, the islands between Shetland and the Scottish mainland, which are about 100 km south of Shetland, a rich, well-farmed, and well-drained agricultural region, far less windy and much more affected by ancient and recent civilizations. Shetland has large areas covered in peat, little of which is suitable for agriculture. There is an aura of isolation about it, one feels a long way from everywhere else. For a biologist, the most important aspect of this isolation is rightly emphasized by Berry and Johnston (1980): flora and fauna are the result of recent colonization, well after the Ice Age.

The islands are renowned for their birds, with huge colonies of sea-birds on the cliffs, from kittiwakes to puffins and guillemots and gannets (*Rissa tridactyla, Fratercula arctica, Uria aalge,* and *Sula bassana*), and many exciting species nesting inland, such as red-throated divers (*Gavia stellata*), skuas (*Stercorarius skua* and *S. parasiticus*), and whimbrels (*Numenius phaeopus*). Otters, or 'dratsies' to the Shetlanders, are abundant; they are slightly smaller than those on mainland Scotland, and with their conspicuous throat patches they are somewhat different in appearance. But they are of the same species, spending their life along sea coasts rather than along lochs and rivers. There are two species of seal in abundance, the common (*Phoca vitulina*) and the grey (*Halichoerus grypus*), both quite important to the otters, and rarely also some other seals, and various whales. Several mammals have been introduced into Shetland, such as stoat (*Mustela erminea*), polecat-ferret (*Mustela furo*), hedgehog (*Erinaceus europaeus*), rabbit (*Oryctolagus cuniculus*), and mountain hare (*Lepus timidus*), and everywhere there are house and wood mice (*Mus musculus* and *Apodemus sylvaticus*). But there are no foxes, badgers, weasels, or brown hares.

The waters separating the islands, the 'Sounds', are mostly less than a kilometre wide, but they have ferocious currents, and many of them must be quite effective barriers even to an otter. Tankers wind their way through these Sounds, past treacherous rocks, past the graves of hundreds of ships which have perished here throughout history. This is the most obvious danger to any marine life in the area, although until now the effects of accidents with tankers have been relatively minor, including even the 1993 disaster of the *Braer*. But one cannot help but fear that a large oil wreck is only a matter of time: Shetland is dominated by oil, by the activities surrounding the huge terminal at Sullom Voe. Tankers will continue to ply these waters for a long time to come, the winds are horrific, the conditions totally unforgiving.

There are other dangers to marine birds and mammals which one sees lurking in the Shetland landscape. Straight pollution for instance, from open rubbish tips spilling straight into the sea, from the many sheep dips discharging into little burns, from food wastes at the floating fish-farms. Persecution of the dratsie, the Shetland otter, was once common, and still there are many of the centuries-old otter traps, 'otter houses', to testify to this. They lie unused now (see p. 255). In the end the most threatening danger to Shetland otters has turned out to be land drainage, and I will discuss this in the following chapters.

We established an intensive study area on a peninsula on the north-east coast of Mainland Shetland, the Lunna Ness (Fig. 1.1, 1.2, 1.3). It is a long finger pointing at the horizon, where the Yell Sound meets the

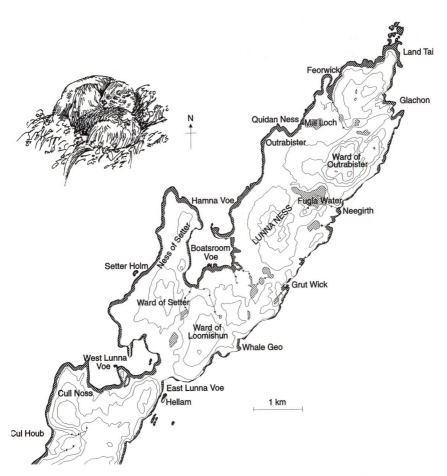

FIG. 1.2 The Lunna Ness study area in Shetland, showing the sheltered bays ('voes') along the north-west where most observations were made, the steep and exposed south-east coast, and small inland lakes.

FIG. 1.3 Prime otter habitat in Shetland: Boatsroom Voe, Lunna. Low, damp coasts with peat.

North Sea. Its total coast was about 30 km, but we used only 16 km really intensively. There were three farms, and a few houses—one of which was our 'field station', close to the shore of one of the large 'voes' (bays), the Boatsroom Voe. There we could sit, myself and the various students and others who helped me, on the seat near the back door, with our back against the white wall, watching otters, seals, and the many birds.

The land of Lunna Ness rises gently from the sea on the north-west side (Fig. 1.3), with steep cliffs along the east coast which is a very rough and exposed shore by any standards; even close inshore the waters go down as deep as 80 m. We concentrated our efforts on the more sheltered, shallow north-west coast, with its three voes separated by more exposed rocky shores, but no cliffs. The Ness is covered with mostly rough, 'natural' grassland and heather, some larger stretches of farmland, and some peat faces where local people cut their fuel. The only livestock was sheep, with a few cattle and ponies. Recently, starting about halfway through our study, another large human influence arrived in the shape of a fish-farm, with its many floating salmon cages in the Boatsroom Voe.

The shore in Shetland is almost invariably narrow, because there is such a small tide; the vertical tidal range is usually only about 1 m. This means that the intertidal zone in the study area is less than 20 m wide (it is usually only a few metres). The water is often very clear, with a visibility much better than anywhere else along the British North Sea shores.

Almost everywhere below the high-tide line is covered in dense algae, down to some 4 to 6 m deep, in a fascinating zonation of different wracks such as the knotted wrack *Ascophyllum nodosum*, bladder wrack *Fucus vesiculosis*, and kelps such as *Laminaria digitata* and *L. hyperborea*. The kelps especially form dense stands, often 2 to 4 m tall, waving forests which appear almost impenetrable to a diving mammal.

There is a great deal of variation in the submerged vegetation of coasts with different exposure to waves. The sheltered voes, for instance, are dominated by knotted wrack and the kelp *Laminaria saccharina*, but more exposed, rough waters are characterized by *L. hyperborea*, the thongweed *Himanthalia elongata,* and *Gigartina stellata*. In fact, the ratio of *L. hyperborea* to *L. saccharina* is used as an index for 'exposure' to wave action (Kitching 1941; Kain 1979), and we have employed it in the description of the habitats for the different prey species of otters (Kruuk *et al.* 1987). There is an abundance of fish, both in species and in numbers (see Chapter 5).

Our freshwater study areas in the north-east of Scotland were various tributaries, sections of main river and lochs in the valleys of the rivers Dee (Fig. 1.4) and Don, which run from the Highlands, reaching the North Sea in Aberdeen (Figs 1.4, 1.5, 1.6). Both are relatively long, fast-

FIG. 1.4 The River Dee in north-east Scotland, one of the study areas.

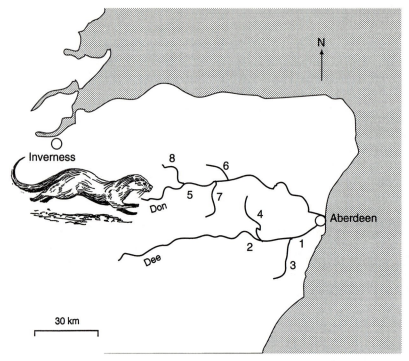

FIG. 1.5 The rivers Dee and Don study areas with tributaries, in north-east Scotland: 1, River Dee, Durris section; 2, River Dee, Woodend section; 3, Sheeoch Burn; 4, Beltie Burn; 5, River Don, Kildrummy section; 6, Esset Burn; 7, Leochel Burn; 8, Mossat Burn.

flowing rivers, in a region of mixed agriculture, forestry, and natural vegetation which is characteristic for a large part of Scotland. Agriculture is mixed pasture and cereal farming in relatively small fields, there are some conifer plantations along the banks, as well as narrow strips of natural woodland here and there, with alder *Alnus glutinosa*, willow *Salix spp.*, and other deciduous trees.

The river bottoms are mostly gravel and boulders, with patches of coarse sand, and very little mud. Both the Dee and the Don are known for their populations of Atlantic salmon *Salmo salar*, and also for sea trout (that is migratory brown trout *S. trutta*) and brown trout, but there are few other species of fish: eels *Anguilla anguilla*, minnows *Phoxinus phoxinus*, three-spined sticklebacks *Gasterosteus aculeatus*, and some brook lamprey *Lampetra planeri*, pike *Esox lucius*, and perch *Perca fluviatilis*. The lochs have almost only eels, perch, and pike. To me, the exciting inhabitants are the fish predators, which are common: the otters, mink *Mustela vison*, the osprey *Pandion haliaetus*, goosanders *Merganser merganser*, grey

FIG. 1.6 Two otters passing a reed bed at Loch Davan, Dee valley, north-east Scotland.

heron *Ardea cinerea*, and the less common great crested grebe *Podiceps cristatus* and cormorant *Phalacrocorax carbo*. These river systems are beautiful, rich areas, with a wealth of wildlife, and I am fortunate to have them right on my doorstep.

METHODS AND FIELD TECHNIQUES

How does one study otters in the field? The many questions which were at the base of our own research obviously needed to be approached from different angles, but an ability to watch the animals, to follow them closely, and record their behaviour was common to almost all. I will discuss some of these general methods here, and more specialized techniques in the relevant chapters.

In Shetland otters were shy, but with caution they could be approached quite closely. The animals are extremely sensitive to smell, so we had to be careful not to get upwind of them, especially when the weather was calm, when they could smell us 200 m away or more. We learned not to show ourselves against the sky-line, and to remain motionless when within sight of an otter. On the other hand, we could sit or lie in full view of the animals, as long as we remained absolutely still, wearing rather dark clothing and sitting against a bank or a rock or whatever. We also learned to make good use of the times when otters were under water, when we could approach closely, or move to a good cover nearby.

In the end much of our observation technique depended on just intuition, on a feel for the animal, where it would emerge, what it would tolerate, and where it would go next. There are no guide books for this, nor is it easy to copy other field workers, because otter watching is best done alone, without a companion to consider. Like many other types of field work with animals, it is something of an art, of which the technical details have to be learned by practice.

The most essential equipment is a pair of binoculars (ideally 7 × 50), and we often used a telescope, with zoom magnification 15–40, mounted on a tripod. The usual distance for behavioural observations of otters was somewhere between 50 and 150 m, on average probably about 70 m, which with binoculars and telescope enabled us to see many details of the animals and what they were doing or eating. A small dictaphone recorded all observations, mumbled softly from underneath the binoculars. Timing was done with an ordinary watch or a small electrical clock; stop-watches were usually too complicated to manipulate at the same time as keeping up with all the otters' antics.

To most people, all otters look alike. However, mature males are visibly larger than females, with broader heads (Figs 1.7, and 1.8), and

Fig. 1.7 Female otter along the Shetland shore. Note difference from the male (Fig. 1.8). (Photograph Thomas Stephan.)

FIG. 1.8 Male otter: note large, broad head.

often swimming more conspicuously, with the tail trailing along the surface. But a youngish male (Fig. 1.9) is difficult to differentiate from a female.

Often there are also clear differences between individuals, and in Shetland we were particularly fortunate because these differences were very marked. The otters always had bright, irregular white or light yellow patches on their throats (Fig. 1.10). Some otters on mainland Britain also have these patches, but they are far less common, and usually much smaller. In itself, the phenomenon that the marking of the Shetland otters is different is interesting: it suggests that there is a degree of isolation between populations. But whatever the further implications, it enabled us to identify otters if we could see them reasonably nearby, and with their throats exposed. Fortunately, otters often show their throats when foraging in the sea, both immediately after they pop up from a dive and when floating on the surface with a small prey, which they eat with their nose pointing sky-wards.

There were, of course, many occasions in our study area when we could not identify an animal: when it was too far away, or just swimming along, or when we only caught a brief glimpse of it. Similarly, in our freshwater studies it was almost never possible to recognize an otter from its markings. Because it was so all-important to recognize individuals, we decided to use coloured ear tags as an additional aid in Shetland. It meant that we had to disturb the animals in order to catch

FIG. 1.9 Young male otter in Shetland.

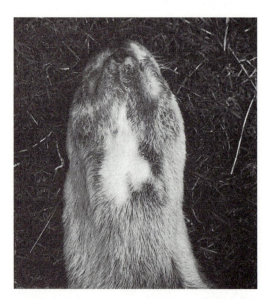

FIG. 1.10 Throat pattern of a Shetland otter.

them, but the tags themselves did not bother the otters in the slightest. Ear tags made them immediately recognizable, for as long as they stayed on, that is for a few months or up to a year.

Using throat patches, ear tags, and other distinguishing marks (white on the upper lips, a missing tail tip, a scar on the nose), we made identity cards for every animal which we met in our Shetland study area, with a name as well as a number if it was a resident. After 5 years we had a total of almost 60 identity cards, of otters which lived in the study area or had lived there, or which had just passed through briefly.

Trapping otters for study was a major concern; it had to be done, but we could not unduly upset them. Previous studies used the 'Hancock trap', originally developed in North America to catch beavers, described in detail by Mitchell-Jones *et al.* (1984). Hancock traps are set on otter trails, buried under sand or vegetation, and a treadle sets off a very strong spring attached to a small 'net', a kind of lid of the trap, made of steel mesh. This lid, with its heavy steel surround, covers the otter with great speed and force, and the animal is squeezed down by the steel netting. It is apparently a very efficient trap, but I found it too violent, and the danger of a sheep, another large animal, or even a person setting it off and breaking a leg seemed unacceptable in the areas where we were working.

Based on the design of the 'otter house', used by the Shetlanders to catch otters when they were still exploited (see p. 256), I developed a trap made of wood, with the dimensions of a large coffin, and put near well-used otter sites. It was smoothed into the landscape with stones or branches, sometimes using just one entrance, sometimes as a two-ended tunnel (Figs 1.11, 1.12). An otter wandering inside would touch a thin pin dangling from the ceiling, which would set off a rat-trap on the top

FIG. 1.11 Trap set to catch otters for radio-tracking in north-east Scotland.

Fig. 1.12 Otter trap in north-east Scotland to catch animals for research. The trap is not baited, but tracks show its attraction to otters.

of the wooden tunnel, and via a couple of strings two nails, which were holding open the doors of the trap would be pulled out and the trap would close. It was something of a Heath-Robinson structure, but it worked.

The traps were left in place for years, not set of course, but open: this was one of the secrets of their success. The otters were used to them, and could walk through the traps without let or hindrance; several times after I took a trap away, I saw otters inspecting the bare site very suspiciously. The traps were never baited with food (otters do not readily take fish which they have not caught themselves, and bait also attracted other animals such as stoats, mink, cats, and hedgehogs), but I often put some fresh spraints inside. When set, the traps were inspected every morning and evening in Shetland, and every morning in mainland Scotland (otters being active mostly during darkness there).

If an otter was caught it was usually fairly quiet, as it was kept in the dark. We immobilized it with an injection of Ketamine, or a mixture of Ketamine and xylazine (Melquist and Hornocker 1979; Mitchell-Jones *et al.* 1984; see also the good review in Reuther 1991). This was administered with a syringe from a blow-pipe, through an inspection cover in the top of the trap (Parish and Kruuk 1982). The animal would then be weighed and measured, ear-tagged or fitted with a radio-transmitter, and put in a box to be released again after recovery.

We followed a few otters by radio-tracking in Shetland (see, for example, Nolet and Kruuk 1989), but in general it was not a success there. In Britain it was then illegal to use internal transmitters in wild animals (this has changed since 1987), and otters do not easily accept anything attached to their body such as a radio-collar or harness. Mitchell-Jones *et al.* (1984) and Green *et al.* (1984) used a radio-harness which stayed on an animal for a few days or weeks, but that was in freshwater. In seawater the animals have much more trouble with a harness, and we now know why (see Chapter 7). There was no doubt that the animals were disturbed by it all the time, showing their distress by frequent rubbing on the grass. We also tried radio-transmitters fixed to loose collars, to collars which were kept in place by some glue to the fur, and small transmitters which were glued onto the fur directly. None of these methods proved acceptable to the otters or to us, the transmitters came off very quickly, and we soon abandoned our radio-tracking efforts in Shetland.

Fortunately, it is now permissible in Britain, as in other countries, to use small internal transmitters, implanting them intraperitoneally as pioneered in the USA (Melquist and Hornocker 1979, 1983), and this has solved the problem for our freshwater studies. The otters do not notice the transmitter, and it is so tiny (the size of an ordinary AA-cel battery) that it does not affect them. The operation is carried out by an experienced veterinarian, surgeon, or physiologist, the transmitter is slipped in when the animal is under anaesthetic, the otter is kept under observation for a week to make sure all is well, and then it is released again at the place of capture.

In freshwater areas, where otters are nocturnal and cover large ranges, radio-tracking is the only way to follow the animals, to discover with whom they associate, where they fish, and which habitats they use. It is now a common technique in all wildlife field studies, with many handbooks to discuss the details. One small extra which we used was that we gave each otter a small dose of a harmless isotope, zinc-65 (Kruuk *et al.* 1980), which is also used in hospitals in medical treatments. This enabled us to recognize the spraints of that individual for about 6 months afterwards: useful for the study of individual food preferences, sprainting habits, range sizes, and the proportion of spraints in areas which came from marked and unmarked otters.

Finally, some of the problems which we faced during field work had to be solved in a more controlled situation, and studies on our captive animals played an important role. An example is the interpretation of data from faecal analysis, where we had information on what was in the spraints and had to translate this into otter diet. This can be done after feeding captive animals known quantities of different fish species, then assessing their 'output' in the spraints. Equally important were the energetics of foraging: in captive animals we could measure oxygen consumption when they were swimming and diving in their pool, and from that we could estimate the costs of foraging in the wild. Details of this are described in the relevant chapters.

THE EURASIAN OTTER (*LUTRA LUTRA* L.)

Lutra lutra lives in watery habitats of many different kinds, in fresh or salt water, in rivers, streams, lakes, marshes, and along sea coasts. It has an enormous geographical range, larger than that of any other species of otter: from the Atlantic seaboard of Europe right across Asia to the Pacific Ocean (including Japan), from northern Lapland and Siberia down to northern Africa, India, and Indonesia. In that distribution there are now conspicuous gaps, where otters have died out recently: in Europe they are virtually gone from most or all of England, France, Germany, Holland, Belgium, Denmark, Sweden, Switzerland, Italy, and parts of other countries (Foster-Turley *et al.* 1990).

Europe has otters along its fringes, therefore, along the Atlantic on the western side, and they are still widespread in the eastern countries. Human civilization has dealt the species mortal blows in Japan, from where it has now all but gone (Foster-Turley *et al.* 1990). However, there is recent evidence of a small recovery of the otter's fortunes in England, where more spraints have been found in the latest surveys than 10 or 20 years ago (Jefferies 1989).

In the British Isles otters are common in Scotland, Wales, and Ireland. Shetland has a healthy population, but there are rather few in Orkney. In most places otters are rarely seen, however, and for evidence of their occurrence one has to rely on tracks (Fig 1.13, 1.14), or on their faeces, the 'spraints' (Fig. 1.15). Detailed descriptions, measurements, and weights of the species in Britain are given in Corbet and Harris (1991).

We are still surprisingly ignorant about the animal, despite a keen interest from every naturalist, and despite the fact that quite a number of books have been written on the natural history and conservation of the otter. These include some very good ones, for instance those by Stephens (1957), Harris (1968), Chanin (1985), and Mason and Macdonald (1986). However, much of the knowledge presented therein comes from captive animals, or from just the occasional glimpses in the wild, and especially from analyses of spraints, their contents, distribution, and numbers. But little has been published on numbers of otters,

FIG. 1.13 Otter tracks in snow, along a stream in north-east Scotland.

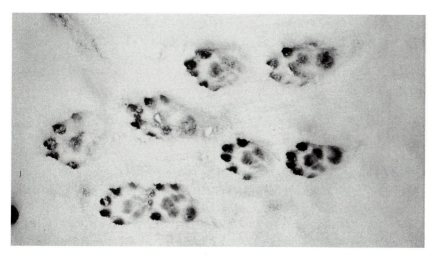

FIG. 1.14 Tracks in snow of an adult female, and a large cub walking on her left.

on their social organization, feeding behaviour, and population dynamics, or even just how and where to watch and study them.

The first and classical research on otter ecology and natural history is that of Sam Erlinge (1967, 1968, 1972), then a PhD student working along the shores of the beautiful lakes of southern Sweden; he was later

FIG. 1.15 Spraint site along a small 'burn' in Shetland.

to become Professor of Ecology in Lund. Many of his observations came
from tracks in the snow as well as from analysis of faeces. He studied
the ranges, territorial behaviour, and food, and his research was a touch-
stone for many, much briefer, studies by others after him. Now, 25 years
on, there is not a single otter to be found in or near his previous study
areas. Much later, in Scotland, Green et al. (1984) were the first to
attach harnesses with radio-transmitters to otters for short periods, and
they obtained further data on the ecology of the species in rivers, generally consistent with what Erlinge had found earlier.

The problem of the decline of the European otter has been addressed
directly by ecologists interested in pollution, summarized by Mason and
Macdonald (1986) in their book on otter conservation, and later by
Mason (1989). The subject is complex, and there is, in fact, no good
hard evidence that any one substance or group of substances is responsible, although there is much circumstantial information.

The bulk of literature on otters deals with their food, with data from
the analysis of spraints (reviews in Chanin 1985; Mason and Macdonald
1986). Otter spraints are easy to find and collect, fish bones are often
passed undamaged, and these bones are relatively easy to identify.
Several authors have analysed thousands of spraints each, especially
from freshwater habitats: in Britain and Ireland especially Wise et al.
(1981), Chanin (1981), and Weir and Bannister (1973, 1978), but also

many others, and more publications are being added every year. There is also a great deal of information from several other European countries. Fish totally dominate the diet everywhere, with eel the single most important species, at least in Britain. However, several other fish have also been found to be important, and rather few studies have been done in fast-flowing rivers, common in mountain areas, where trout and salmon predominate (Jenkins and Harper 1980; Wise *et al.* 1981; Chanin 1981; Kruuk *et al.* 1993).

There is also published information on the food of otters in the sea, from Shetland, and from the Norwegian coast (Watson 1978; Herfst 1984; Lightfoot, in Mason and Macdonald 1986). There the animals ate mostly small, bottom-living fish such as rocklings, eelpout, butterfish, and flatfish, as well as crabs.

Spraints not only provide the data on which most of our knowledge of food is founded, but they have been used extensively also to establish the presence of otters and to assess increases or decreases in numbers and distribution and the use of different habitat types. National otter surveys are basically spraint surveys (see, for example, Lenton *et al.* 1980; Green and Green 1980; Chapman and Chapman 1982), and most of our knowledge of otter decline comes from finding fewer spraints. However, there are problems with such use of 'spraint indices', which are discussed in Chapter 8.

OTHER OTTERS, AND OTHER MUSTELIDS

Otters belong to the subfamily Lutrinae, of the family Mustelidae. The Mustelidae are the largest of the seven families of Carnivora, with about 60 species (Ewer 1973; Macdonald 1984); other subfamilies are that of the martens and weasels, the badgers, the honey-badger, and the skunks.

There are 13 species of otter, of Lutrinae. In Europe there is only the one, variously called the otter, or European otter or Eurasian otter: *Lutra lutra*. North America has the river otter *Lutra canadensis*, which is in many ways similar to the European one, but there are important differences: for instance in that the river otter has a delayed implantation, unlike the Eurasian one. This means that its gestation period is prolonged, as the fetus does not start growing until sometimes several months after fertilization.

North America also has the sea otter *Enhydra lutra*, along the Pacific coast, which is totally different from all other otters. It is the most aquatic of carnivores, even more so than seals; it eats, sleeps, mates, gives birth, and rears its cubs in the sea. Several excellent studies have been made on the North American otters, such as a classical, large-scale radio-tracking project on the river otter in Idaho by Melquist and Hornocker (1983). For sea otters the monograph by Kenyon (1959) will stand for generations, having been the basis of much first-rate research (for example by Estes and his colleagues (Estes *et al.* 1982; Estes and Van Blaricom 1985; Estes 1989)). There is a good review of sea otter studies by Riedman and Estes (1990).

The two North American otter species are especially relevant to our work in Shetland, as they sometimes (river otter) or always (sea otter) also live in a marine habitat. Similarly, the sea cat *Lutra felina*, a small otter from the southern Pacific coast of South America, has to cope with the specifically marine problems which I will discuss later. This species occurs only in seawater (Ostfeld *et al.* 1989), in an arid climate.

There are three other species of otter in South America, but they use freshwater only; the best-known is the giant otter *Pteronura brasiliensis* (Duplaix 1980), and there are the southern river otter *L. provocax* and the neotropical otter *L. longicaudis*, both very little studied and virtually unknown. The giant otter grows up to 2 m long, and lives in noisy family parties of up to 20 animals (most commonly fewer than 10), in rivers and lakes of the tropical rainforests.

In Africa there are four otters, of which *L. lutra* occurs only in a small area north of the Sahara. *L. maculicollis*, the spotted-necked otter, is a small, freshwater species from south of the Sahara (Procter 1963; Kruuk and Goudswaard 1990). Of the two large clawless otters, *Aonyx capensis* and *A. congica*, the first, the Cape clawless, uses rivers and lakes, but also sea water, and it has been studied more intensively along sea coasts than elsewhere (Rowe-Rowe 1977; van der Zee 1981; Arden-Clarke 1986; Skinner and Smithers 1990). Perhaps it is easier to watch along coasts, as are our Shetland otters. It is larger than *L. lutra,* and much more a crab-eater. *A. congica* is almost totally unknown.

In Asia there are five species: *L. lutra*, almost throughout the continent, the sea otter *Enhydra lutris* along the northern shores of the Pacific Ocean (in a continuation from its North American distribution), the small-clawed otter *Aonyx cinerea* (a small, mostly crab-eating species), the smooth otter *L. perspicillata* (larger than the Eurasian otter, eating mostly fish), and the hairy-nosed otter *L. sumatrana* (about the same size as the Eurasian one, but nothing is known about its habits). The last three species occur mostly in fresh water, but also along mangrove coasts. In some areas of south-east Asia as many as three different species can be found together.

In many ways all these species are remarkably similar, they look the same and behave in the same way, they are otters without the slightest doubt. The spotted-necked otters which I watched in Lake Victoria were fascinating to me because of the differences in detail, which I detected on comparing them with Eurasian otters: but even more astonishing was the tremendous likeness between the two species. Similarly, the three which I studied in the river Huai Kha Khaeng in Thailand (common, smooth, and small-clawed otter) were amazingly alike in many details (Kruuk *et al.* 1993, 1994).

However, despite the apparent uniformity, and despite the effort which has been put into studies of otter ecology by so many researchers, we still know relatively little about the animals in general, with the exception of the sea otter. The reason for this is not lack of directed effort or imagination of the scientists, but, in apparent contradiction to what I said above, the large variability which faces them: not differences between species, but especially the variation in ecology and behaviour within species, between populations and individuals. Furthermore, otters are difficult to observe, often nocturnal, secretive, and in habitats which are not easy to get at.

Somehow, we will have to come to grips with the variability and quantify the extent of variation, the ecological and behavioural limits for each species. It is a prerequisite if we want to get some idea of how they evolved, what kinds of environmental pressures they are able to withstand, and how they are going to continue in the face of future changes in their ecology owing to the influence of man.

2

Spatial organization, holts, and habitat

The Eurasian otter may be solitary, or it may live in group territories in the sea or fresh water. Female otters in Shetland were living in group ranges of 5–14 km of coast, with little overlap between ranges, and ranges were defended against females from outside. Individual females each had their own core area within the group range. In rivers they were more solitary, but in freshwater lakes females shared territories. Females within group ranges were probably related to each other. Males were larger, they had larger ranges, overlapping with several female group areas, and individual males overlapped widely with each other, although they were aggressively territorial. Males had a different habitat preference, living along more exposed sea coasts or in larger rivers than females, and in Shetland they entered female areas especially in late spring, the mating season. Transient otters appeared to have no specific limits to their home range.

Habitat characteristics are described for otters along sea coasts as well as inland. The dominating requirement along coasts is the presence of fresh water, in streams, pools, or underground. In freshwater lakes reed beds are very important, along rivers otters show preference for places where fish can be easily caught (for example under banks and in riffles), but bankside vegetation is largely unimportant to them. Holts are made close to water except natal holts, which may be far away. In many freshwater areas otters only rarely use holts, and they may build covered couches.

INTRODUCTION

In carnivores, more than in most other mammals, there is a large variety of spacing patterns, of many different social systems and dispersions—but there are also similarities. Almost all carnivores are territorial, defending an area against others of the same species (Ewer 1973), even if only against particular categories of conspecifics. Usually such an area has a clear and well-defined border, and the 'owner' reinforces territorial claims with aggression, and scent marks of some kind. In many species such territories are inhabited by almost totally solitary individuals, but there are also group-living carnivores, living in small, tight packs, or in widely dispersed groups, defending the borders of their own group territory against other packs, or clans, or prides, or whatever their group system is called. In short, we recognize a simple distinction between group-living and solitary species (Sandell 1989).

Knowledge of spatial organization is important in order to model the relationship between, on the one hand, numbers and densities of the predator, and on the other hand the dispersion of its food, shelter, and other requirements. It is likely that spacing patterns are affected by resource availability and utilization (Macdonald 1983). These spacing patterns will have to be quantified in order to understand the underlying mechanisms such as competition, or reproductive suppression (Creel and Creel 1991), which are important aspects of the regulation of animal numbers.

Also, and on a more practical level, there is a need to know the extent of a species' variability in spacing, in dispersal patterns, and in underlying causes such as resource distribution. For conservation we need to know this as well as the overall limiting factors to a species' numbers if we are ever going to effectively manage these animals. This need is urgent, especially for conservation priorities such as the otter.

The organization of animal societies into groups and group territories is particularly fascinating, especially when we start to quantify these social systems, the phenomena of communities of different sizes living in areas of varying dimensions. Often members of such communities cooperate in one way or another, for example in hunting (Kruuk 1975), rearing offspring (Macdonald and Moehlman 1983), defence against predation (Rood 1986), or in other ways (Sandell 1989).

Frequently there is little or no correlation between the size of the group and the area it occupies: individuals of the same species may live alone or in a group of almost any size, in small territories or in relatively huge ones, as for instance in badgers, hyaenas, wolves, or foxes (Mech 1970; Kruuk 1972, 1978; Macdonald 1980b, 1983; Kruuk and Macdonald

1985; Mills 1990). There have been several studies to explain this variation, asking why some individuals in some species disperse whilst others stay in their natal range and add to the group, what it is in the environment that sets limits to groups or the size of their territories, and how the structure of these societies is adapted to their environment.

Much of such variation in spatial organization within a species can be explained by the distribution and richness of the resources. This has been formalized for carnivores by David Macdonald, in his Resource Dispersion Hypothesis (MacDonald 1983). I have discussed it in some detail in *The Social Badger* (Kruuk 1989a), and I will come back to it in Chapter 9. One could expect the relation between resources and territory size to be relatively simple in the 'solitary' species (Sandell 1989), uncluttered as they are by variations in group size; however, there are not many studies of solitary carnivores in which the dispersion of resources has been documented. One of the aims of our otter project was to do this for such a solitary animal.

In the case of the otter the expected simplicity turned out to be highly complicated (Kruuk and Moorhouse 1991). This was partly because what we thought to be a solitary species (Erlinge 1967; Chanin 1985; Mason and Macdonald 1986), was not, and as I will show, the distinction between 'solitary' and 'social' has little meaning in cases such as that of the otter where the terms do no more than indicate the extremes of a spectrum of organizations. We first noticed this in Shetland, where otters live neither singly nor do they exist in clear groups, and where no single term suffices to describe the degree of their gregariousness.

ORGANIZATION IN MARINE HABITATS

The Shetland study: introduction

When we first started our intensive project on Shetland we were advised by several people there, such as Bobby Tulloch and Peter Dale, who had been watching otters themselves for years, and who directed us to the right places. Some coasts are much better for otters than others, and they have remained so over the years that we worked there: a clear indication of some pronounced habitat preference. This we tried to quantify, as I describe in the following section, but our main interest was to understand the relationships between otters, the organization of individual ranges, systems of spacing, and groupings. The study area which we set up in one of the best, high-density otter areas in Shetland was ideal for this purpose, with good possibilities of observation and recognition of individuals, but especially in the beginning this only added to a sense of frustration at being unable to see any pattern in the movements of the otters. Previous studies by others were of little help, and we were disadvantaged in being able to do only very limited radio-tracking. It took several years along those Shetland coasts before a recognizable pattern of spatial organization emerged, a pattern quite different from what was known in any other carnivore.

Habitat selection in Shetland

To find where otters occurred in Shetland, we surveyed the coasts of all the large islands and many of the small ones (Kruuk *et al.* 1989). The main purpose of the survey was to estimate the total population of otters (Chapter 9) after we had already studied the animals intensively in one study area. Besides this we also obtained information on differences in otter numbers along the various types of coast. We divided the 1350 km shore line into 5 km sections and a number of small islands, and classified those as:

(1)   sections dominated by steep cliffs (more than 10 m high); number of sections $(n) = 89$;

(2)   shores along agriculture ('improved' soils), $n = 61$;

(3)   shores along peat, $n = 40$;

(4)   other types, $n = 47$;

(5)   built-up areas, $n = 5$;

(6)   38 small islands (often peat).

Sections were numbered, and the ones we surveyed (about one-third in all) were chosen randomly from the different types of coast. In each of those we searched for otter holts, as it had been suggested that there was a correlation between otter numbers and numbers of holts along the shore (Conroy and French 1987). This searching of coasts consisted of walking, zigzag, through a 100 m wide strip, inspecting all likely sites—a relatively easy procedure, as the holts were generally conspicuous, the terrain was fairly easy to walk, and there was no tall vegetation.

There were large differences in holt density (Fig. 2.1), with the highest numbers along peaty shores (Figs 1.3, 2.2). In these peaty areas we found a mean of 13.3 holts per 5 km, ± 7.7 s.e., $n = 22$, in which s.e. is the standard error of the estimate and n is the number of sections which we counted. On the 12 small islands there were fewer (7.8 holts ± 4.1 per 5 km); cliffs and agricultural coasts had only 1.8 ± 2.3 and 1.6 ± 2.6 holts ($n = 19$ and 21), other coasts were intermediate with 4.1 ± 3.9 holts ($n = 21$), and there were no holts in built-up areas (close to housing). Coasts such as the whole southern part of Mainland, where the 1993 oil disaster from the stranding of the *Braer* took place, had a holt density of only 0.8 per 5 km even before the accident, the lowest in Shetland. This variation was highly significant (one-way analysis of variance $F = 28.79$, $p < 0.001$), and demonstrated the clear preference of otters for the gently sloping, peaty coasts of Shetland, without agriculture, and without cliffs. The habitat preference was also evident when we looked at the 5 km coastal sections in greater detail, dividing them into 1 km stretches and estimating the different types of coast within each one (Kruuk *et al.* 1989). I will show later (Chapter 7) that the otters' preference for particular stretches of Shetland coast is likely to be based on their need for fresh water.

Spatial organization in Shetland

General

Our main study area was about 30 km of coast of Lunna Ness, a peninsula of the north-east of the largest island, Mainland (see Fig. 1.2, p. 9). Most of the observations were made along 16 km of this, on the north-west side of the peninsula, which is a relatively shallow coast with gently sloping rocks and gravels, covered in algae below the high water mark

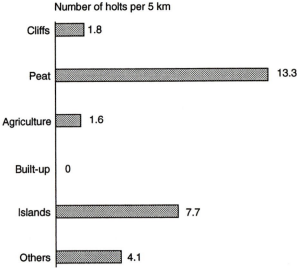

FIG. 2.1 Numbers of otter holts along the Shetland coasts, in different habitats. Densities shown are the number of holts per 5 km section, in a strip along the sea of 100 m width.

FIG. 2.2 Main holt along Shetland shore. Many entrances, smoothed rolling areas, spraints on stones and grass.

and by peaty soils and grasslands above it. The water is rarely more than 8–10 m deep at 100 m from the shore. Most of the north-west coast of the peninsula is fairly exposed to the waves and the interminable winds, but there are three bays, the Boatsroom Voe, West Lunna Voe, and Feorwick which are sheltered havens for otters and observers alike. The east side of the peninsula has many steep cliffs and slopes, and the water is much deeper there.

Although my students and I lived along the shore of one of the bays in the middle of the study area, and although we knew many of the otters individually by sight, it was several years into our project before a pattern began to emerge in the social relationships and the way in which the animals used the area (Kruuk and Moorhouse 1991).

We came to realize that some females were always there, several of them in the same area for at least a couple of years, whilst others were only transitory. Observations of individual males were much more erratic than those of females, they obviously used much larger areas of coast than the females. Interestingly, the females who were 'residents' did not go beyond particular points along the coast, but though always alone or with their own cubs, they did share their ranges with other females, and there were no clear, simple territories.

There was, on average, about one adult otter for each kilometre of coast in the study area (Chapter 8), and every one used a stretch of shore several kilometres long. To describe the organization of this, the animals were divided into three distinct categories: resident females, resident males, and transient animals. The most important category, around which the whole system revolved, was that of the resident females.

All otters used only a fairly narrow strip of water and land along the coast. Occasionally they moved outside this, of course, but the vast majority of our observations were made close to the line where land meets water. For instance, in a sample of 500 dives which we saw otters make whilst they swam along the coast, in March and July 1986, we estimated the distance of the dive site from the shore and found that 62 per cent were within 20 m of land, 84 per cent were within 50 m, and 98 per cent within 80 m. Also on land otters were usually close to the shore line; Moorhouse (1988) measured the distance of otter holts from the sea in the study area, and found 77 per cent of 112 holts within 100 m. Some holts were much further inland, and they were very important to the otters, but during the day otters would almost invariably curl up and sleep in the seaweed or amongst the rocks along the shore.

All this was reason why usually we talked of otter ranges in terms of length of coast used, not size of area as one does for most other mammals, and previous authors have done the same (Erlinge 1967, 1968; Melquist and Hornocker 1983; Green *et al.* 1984). However, for some purposes this is clearly not correct, because the width of the strip of coast used by otters varied with the slope of the coast, the intensity of use of the strip of water declines with depth (p. 171), and because also in fresh water the width of the stream appears to be very important (p. 54). Nevertheless, just for the purpose of describing the spatial organization I will maintain the use of that much easier measure, the length of coast. For easy recording into the field we divided the coast of the study area into sections, each about 350 m long, and defined in the field by natural features such as rocks, small streams, or whatever (Fig. 2.3).

Ranges of resident females

One problem with establishing how large an area an animal uses is that with increased numbers of observations the 'known home range' gets larger. We plotted the size of the known range against the time that the animal had been observed (Fig. 2.4); in a resident animal the resulting curve reaches a plateau, an asymptote, and this indicates the extent of the known home range (Woollard and Harris 1990, and previous authors). In an itinerant, roving, non-resident animal there will be no

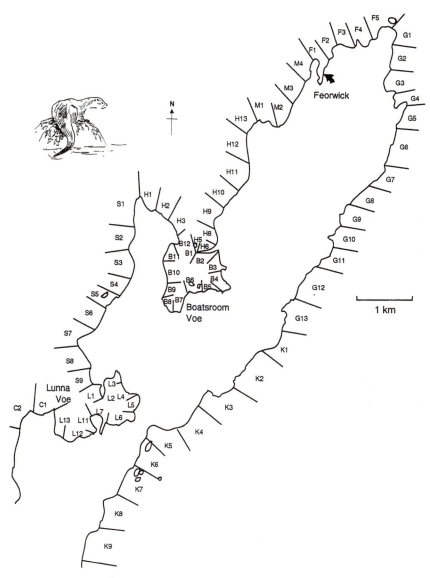

FIG. 2.3 Coast of the Lunna Ness study area in Shetland, divided into observation sections used for reference in the field to establish home range sizes as in Figs 2.4, 2.5 and 2.7.

asymptote, and the known range will continue to increase with observation time. Fig. 2.4 shows the size of ranges of five female otters accumulated over different lengths of time, animals for which we had a relatively large amount of data. For three of the otters, with the largest

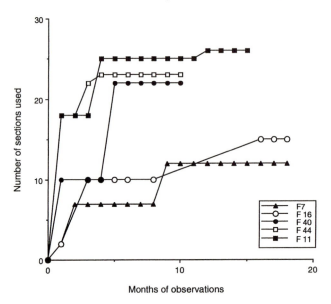

FIG. 2.4 How long to observe an otter, to estimate the size of the home range (Shetland). Vertical: the size of the known home range of some of the females in the Shetland study area (expressed as numbers of sections of coast). Horizontal: the length of time we had followed them. For most animals, the known range size did not increase substantially after 3 months, often much less. Each graph refers to one female. (After Kruuk and Moorhouse 1991.)

ranges, the curve reached an asymptote after about 4 months of observation. For the other two an initial asymptote was reached after about 2 months, and then much later the area which the animal used expanded again, in both cases probably associated with a new litter of cubs. We decided that for the purpose of comparing range sizes between individual otters we should only use observations of animals for which we had information from at least 10 observations, spread over at least 4 months. The shape of the curves in Fig. 2.4 also suggests that within the home range the *core area*, that is the smallest part of the range where an animal spends at least 50 per cent of its time, would often be apparent with far fewer observations, within days or a few weeks.

Over 5 years we recorded movements and observations of many different females through the numbered sections of the coast of Lunna Ness, and for 15 of them there was sufficient information to get an idea of their home range size. Fig. 2.5 shows the ranges of a number of otters as straight lines; these individuals did not all use the area at the same time, and simultaneously with them there were several other otters present for which we did not get enough observations, especially at the

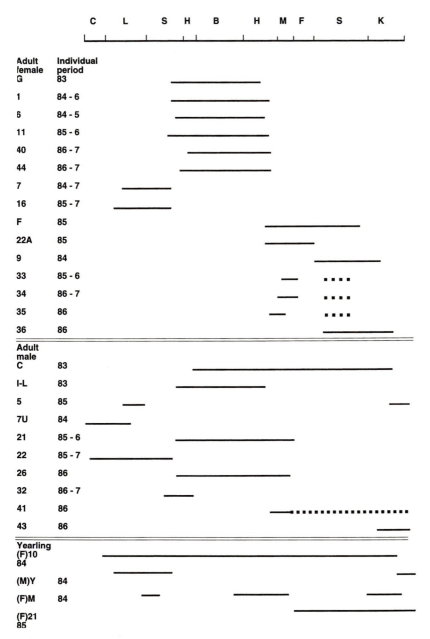

FIG. 2.5 The size of ranges of individual otters in Shetland. Area as in Fig. 2.3. Each line refers to one individual, and the year(s) in which it was observed. Dotted lines = presence inferred. Females, males, and yearling otters are grouped separately. Note that not all the animals used the area simultaneously. (After Kruuk and Moorhouse 1991.)

beginning of our study. For the adult females there were three boundary areas, therefore four distinct ranges, each being used by several animals (although we obtained almost no data on the animals in the southern-most range).

Despite the fact that the final picture was not complete, it was obvious that females living along a particular coast recognized the same bound-aries, not only when they were there simultaneously but also during con-secutive years. The exact location of these boundaries was found by marking turning-around points when otters were swimming along the coast, as well as from mapped sightings. Between Cul and Lunna ranges, the border was a stretch of rocky beach, the Lunna and Boatsroom females had a small stone dyke sticking out into the sea to indicate where their ranges ended, and between the Boatsroom and North-East ranges the boundary was a small stream entering the sea. We therefore knew the length of two of the ranges with some certainty: Lunna was about 4.7 km long and Boatsrooom 6.4 km. The North-East range, stretching along the cliffs and with deep water, was possibly about 14 km long, judging from the movements of otters within it in many casual observa-tions. However, some females in that area probably had ranges smaller than 14 km, but the overall North-East range of otter females was cer-tainly much longer (and narrower) than the others, with otters staying very closely inshore because of deeper water. The boundaries between female ranges appeared to remain in the same place over the 5 years that we were there, with a successsion of different otters involved.

Because at any one time there were several otters living in each of the ranges in the study area, we called them group ranges. The Lunna range was used by two females, the Boatsroom range by up to four simultaneously, the North-East range by at least five. They were resi-dent as adults for up to 2 years, then they disappeared and may have died.

Within each home range, individual females did not use all parts of the shore with the same intensity, and each appeared to have her own preferred stretches of coast. This led to the definition of the *core area*, which was for each female the 350 m section where she was seen most often, plus the adjoining sections which made up the shortest stretch of coast where she was observed at least 50 per cent of the time. This definition followed the study of core areas in badgers by Latour (1988), after a method developed by Samuel *et al.* (1985), and it enabled me to compare the range use system in badgers and otters (Chapter 9).

The system of core areas in the Lunna and Boatsroom home ranges is summarized in Fig. 2.6 and 2.7. The results showed that although several females could have identical home ranges their core areas were quite separate. Interestingly, the core areas of some females changed

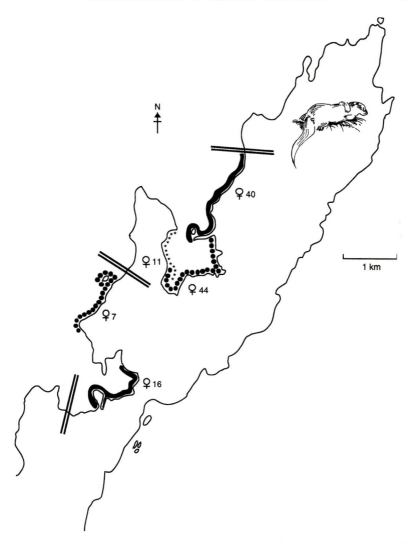

FIG. 2.6 The total ranges and core areas of females July 1986–June 1987; see also Fig. 2.7. Boundaries between group ranges are shown as double lines, core areas have different symbols for each otter. Each female used the whole coast between the double lines (her home range), but spent most (>50 per cent) of her time in her core area.

from one year to the next, although the home range remained the same. For instance, the mother and daughter inhabiting Lunna Voe and nearby coasts had exactly the same core area during the daughter's first independent year of life, meeting frequently and often foraging together. But a year later the daughter moved to another part of the range, from where the previous female occupant had disappeared;

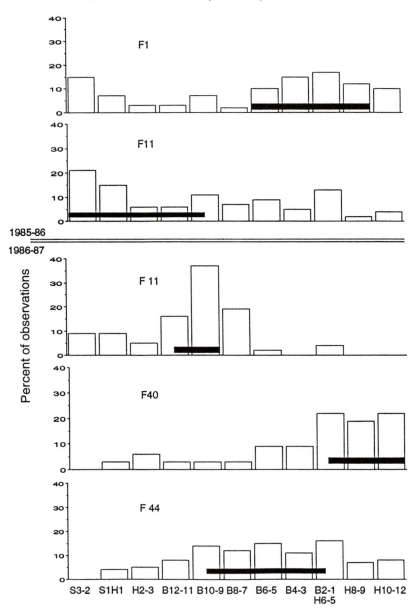

FIG. 2.7a. Group ranges and core areas of otters in Shetland. Shown are the proportions of time spent by different otters in given groups of coastal sections. The core area (the shortest stretch where each animal spent more than 50 per cent of her time) is emphasized. Top: Boatsroom Voe range July 1985–June 1986: females F1 and F11 both used the whole range, but occupied separate core areas. Bottom: Boatsroom Voe range July 1986–June 1987: one of the previous years' females and two others, with separate core areas.

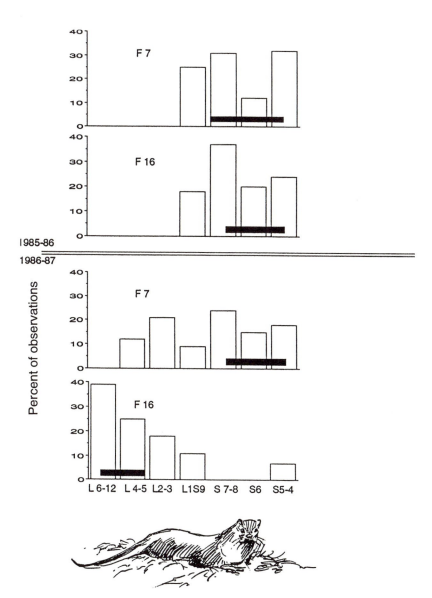

FIG. 2.7b. Top : Lunna Voe range, July 1985–June 1986: mother (female F7) and her independent daughter from age 1 to 2 (female F16). Bottom: same Lunna range, same females but 1 year later, each now rearing a litter of cubs. (After Kruuk and Moorhouse 1991.)

FIG. 2.8 Female F16 with large cub, in her core area at Lunna Voe, Shetland.

mother and daughter still occasionally visited each other's sections, and both produced a litter of cubs (Fig. 2.8). The mother lost hers, however, and she herself disappeared soon after. In the neighbouring Boatsroom Voe we observed a female which had cubs in one core area one year, then changed to another core area for the next year where she had cubs again; the previous occupant had disappeared (Fig. 2.7). In the long North-East range we had far fewer observations and there were more female inhabitants; however, results there also supported the suggestion of a group range system with individual core areas.

In general, females living in the same home range avoided each other, resulting in a division of the total home range into individual core areas, but at the same time there was a great deal of mutual tolerance. This was quite different from the relationships with females from outside the group home range; they appeared to be aggressively and effectively excluded (see Chapter 3). The female group ranges were, in fact, intrasexual group territories, defended areas.

Ranges of resident males

Male ('dog') otters were seen less often than females ('bitches'), and when we followed them they moved over much larger distances. Their somewhat different behaviour meant that it was difficult to get enough observations on individual males to recognize their total ranges. Also there were fewer of them, at least in our main study area. For instance, in 1985 and 1986, in a sample of 887 observations of adults when we could be certain of the sex of the otter, only 34 per cent were males, 66 per cent females. Secondly, males occurred more often along the more exposed coasts, where they could be overlooked more easily, rather than in the sheltered bays, the voes. Along the wild, rocky eastern shores of the Lunna peninsula, 39 per cent of the adult otters we saw were males, but in the voes as few as 28 per cent, a difference which was statistically highly significant ($\chi^2 = 11.4, p < 0.001$) (Kruuk and Moorhouse 1990). In both these habitats there was a strong seasonal trend (Fig. 2.9): during April and May (the mating season, see Chapter 3) 61 per cent of otter observations were of males, against only 7 per cent in December, again a highly significant difference (runs test, $p < 0.005$; Kruuk and Moorhouse 1991).

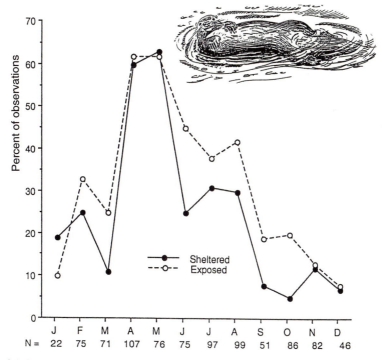

FIG. 2.9 Proportion of males amongst otters in Lunna, at different times of the year. Each point shows the percentage of observations during that month in which otters were males. N = total numbers of otters observed. There were many more males in late spring, and more along the exposed coasts. Both seasonality and difference between exposed and sheltered areas are statistically significant (runs test $p < 0.05$, and $\chi^2 = 11.4, p < 0.001$). (After Kruuk and Moorhouse 1991.)

Dog otters are quite considerably larger than bitches, as in other mustelids (Powell 1979). In Shetland, for instance, the mean weight of males more than 2 years old was 7.35 kg (95 per cent confidence limits ± 0.46 kg; number of animals weighed $n = 31$), and of females 5.05 kg (± 0.29 kg; $n = 42$). These body weights, incidentally, are much lower than those of otters from the British mainland, where an average weight of 10.1 kg is recorded for males, 7.0 kg for females (Chanin 1991). Shetland males were almost 1.5 times heavier than females, therefore, and one could expect larger home ranges on that basis alone, since home range size in many species is correlated with body weight (Clutton-Brock and Harvey 1977).

Taken together, our observations suggested that dogs had much larger home ranges than bitches, they usually lived along more exposed coasts, and seasonally entered the female areas. Even when they were present in our study area, their ranges were relatively large (Fig. 2.5), up to 19.3 km, but there was not one single male for which we knew the full extent of his movements. If we had been able to follow males with radio-tracking over extended periods it would have answered many questions, but they entered our traps less often than females did, and any trans-mitters we managed to attach to males never stayed on for long enough. It was only after the end of our Shetland study that it became legally possible to use internal implantable transmitters for otters in Britain, the only method which is really suitable (Melquist and Hornocker 1979).

The boundaries between the female ranges in Lunna and Boatsroom appeared to be observed also by at least several of the males, which turned back at exactly the same point along the coast. But the data sug-gested that the boundary between the Boatsroom and North-East females (Fig. 2.6) was ignored by males; they always travelled on there without let or hindrance. The ranges of individual males overlapped with at least two groups of females, and probably often with more. Each male range was used by several dog otters simultaneously, but the pattern of overlap between them was unclear, and much more informa-tion is needed. Whatever the range-sharing pattern between males, they were almost invariably aggressive to one another, much more so than females (see Chapter 3), and there appeared to be some aggressive, intrasexual territorial system underlying the dispersion, as in many other mustelids (Powell 1979).

Ranges of transient otters

This category of animals was better studied in Shetland than were the resident dog otters, because for some reason they were more likely to walk into our box-traps, and we were able to get some good radio-track-ing data on four of them, three females and one male (Fig. 2.5). One of the transients was followed over 28 km of coast, another moved over at least 40 km. One female disappeared immediately after release, and she was discovered 3 months later on the island of Whalsay, a distance of only 15 km as the crow flies, but much further when measured along the winding shore line, which would have been her route.

Following these transient otters along the shores I noticed that they were often doing something new or unexpected, behaving as if explor-ing, quite differently from our regular residents. An otter was not recorded as a transient until it had been followed for a few days, when we realized that it did not stay in one particular area but just kept on en-larging its range, moving further and further afield. Characteristically, such transients used rather inferior types of holts, away from the usual otter haunts, often rabbit burrows with nothing to show from the outside that there was an otter in it: only the radio-transmitter told the story. Their food, too, was often rather inferior compared with that of the residents; for instance one female transient specialized in catching rabbits in their burrows, and some others would eat mostly crabs. For further details see Chapter 4.

Holts along the Shetland coast

Within the system of home ranges along the Shetland shores the 'holts' of the otters, the dens, play an obviously very important role, much more so than in other, freshwater areas (p. 59). Otters use them to sleep in at night, but they are essential also for other purposes. Little was known about these holts along sea shores apart from a few casual obser-vations (Harvie-Brown and Buckley 1892; Kruuk and Hewson 1978), until Andrew Moorhouse did a very detailed analysis of their structure, location, spacing, and use in our Shetland study area (Moorhouse 1988). He found that otters usually constructed their own holts, digging exten-sive systems of tunnels and chambers, and providing them with bedding, although sometimes they also used existing caves and tunnels, or rabbit warrens.

Moorhouse excavated nine holts in the Shetland peat layers, reinstat-ing them again afterwards, and he measured and mapped the system of tunnels and chambers (Fig. 2.10). Some of these holts were used by the otters as natal holts, where females gave birth to cubs (see p. 85), but Moorhouse only excavated them at least a month after the family had left in order not to disturb them; once a natal holt was abandoned, the otters rarely returned until the following year. Main entrances to all

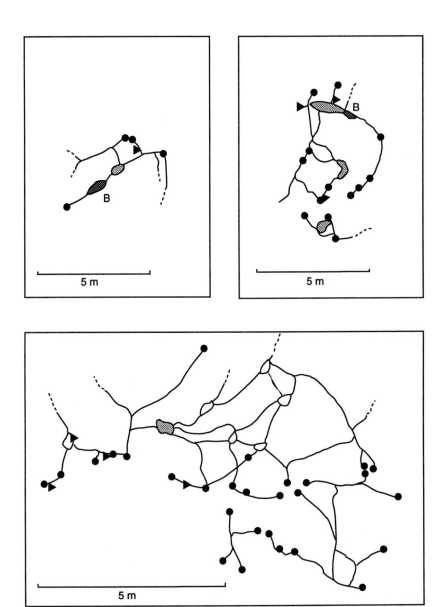

FIG. 2.10 Some plans of otter holts in Shetland. ● = entrance, ▼ = underground latrine; shaded areas are sleeping chambers with bedding, bath chambers with water are marked with B. The bottom holt was abandoned at the time of excavation, and no fresh water was present. (From Moorhouse 1988.)

otter holts have a characteristic shape, being larger and also wider than high (on average 27 cm wide, 17 cm high). This is different from the burrows of rabbits, which are also common in Shetland; they have a circular entrance. Regularly used otter holts also show a characteristic smoothly worn entrance, and often one finds tracks and/or faeces ('spraints'; Figs 2.2, 2.11).

The total length of tunnelling inside the holt may be quite considerable, in Moorhouse's sample up to 51 m, but several holts we saw must have been much larger still. The usual length of tunnel was between 10 and 20 m, extending horizontally at a depth of commonly about 0.5 m under the surface. The chambers could be just widened parts of a tunnel, or at the end of special side-tunnels, and they were lined with bedding, often masses of it. Otters used heather (*Calluna vulgaris*) for this, or grasses, and holts close to the shore line were often lined with fresh seaweeds, especially knotted wrack (*Ascophyllum nodosum*). Several times we watched otters taking bedding in, after first biting off the vegetation, then carrying it into the holt in their mouth; on one occasion Moorhouse observed a male dragging bedding in by taking it under the chin and on his fore legs, walking backwards into the holt entrance— exactly the way in which badgers (*Meles meles*) always do (Neal 1986; Kruuk 1989*a*). Many of the chambers in excavated holts had large sheets of plastic as bedding, such as the fertilizer bags which litter the

FIG. 2.11 Holt on hill top, near the shore of Lunna Voe, Shetland. Note conspicuous spraints with crab remains, near the entrance.

Shetland shores, and I saw otters drag them in. We may object to this horrible evidence of civilization along the beaches, but otters took to the plastic culture with gusto.

All the holts excavated by Moorhouse, which were used by otters at the time, had pools of fresh water inside, often quite far from the entrance and invisible from the outside. These pools could be connected with small underground streams, or they could contain water which collected in a hollow in an impermeable layer below the peat, and they were very important as built-in bathrooms (see p. 198). Moorhouse's observations strongly suggested that otters abandoned holts in which the water dried up.

The Lunna study area had large numbers of otter holts, but Moorhouse also studied several other coasts, including small islands. For these surveys only those holts were recorded which were in active use by otters at the time, that is where tracks and/or spraints were found at the entrance, and where the soil looked recently worn. If entrances were more than 10 m apart and not obviously connected underground they were recorded as separate holts. On Lunna alone there were well over 100 such used holts along 16 km of shore and inland, many in every otter home range, and this was found wherever there were otters on Shetland. The main aim of these surveys was to establish patterns in the distribution of the holts, and to investigate a possible relationship between numbers of holts and numbers of otters.

Most holts were found close to the sea coast; on Lunna 77 per cent of the 112 known holts were within 100 m of the shore line, and only 13 per cent were more than 200 m inland. The inland holts were more difficult to find, so Moorhouse focused on holts within 100 m of the sea. There was a significant difference from random in the distribution; the holts were clearly clustered, and in almost all areas they were at or near sources of fresh water—often clearly 'wet', with muddy entrances, or with water visible within. In the large Shetland otter survey we found that 61 per cent of 1 km coastal sections in which there was an otter holt also contained open fresh water, as compared with only 34 per cent for sections without otter holts, a highly significant difference (Fig. 2.12) (Kruuk *et al.* 1989). There was a close association between holts and wet, peaty coasts (Fig. 2.1, p. 30); small islands, too, were invariably inhabited by otters provided that there were sources of fresh water present. It appeared that this association of holts with fresh water (on the surface, or underground) was an important reason why we found so few signs of otters near the well-drained agricultural areas.

In every home range of otter females there were many holts actively used by the animals, and we often saw them go from one holt to another, especially when they were accompanied by cubs. There was a clear correlation between the number of resident females inhabiting an area and the number of active holts there; using observations from the Lunna study area as well as several small islands (Moorhouse 1988) we found a good linear relationship (Fig. 2.13), which showed that there were about three times as many active holts as there were resident females.

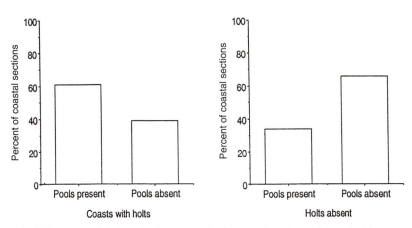

FIG. 2.12 Presence of obvious freshwater pools, in coastal sections with and without otter holts ($\chi^2 = 33.5, p < 0.001$).

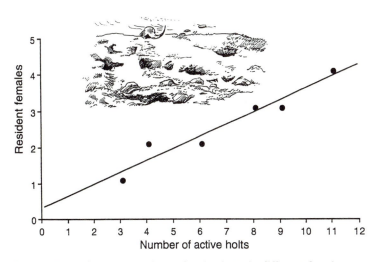

FIG. 2.13 Correlation between numbers of active holts in different female ranges, and the number of resident females present in those ranges: $r = 0.97$, $r^2 = 0.93$, $p < 0.001$; $y = 0.24 + 0.33x$. (From Kruuk *et al.* 1989.)

Moorhouse distinguished 'main holts', 'subsidiary holts', and an intermediate category, based on frequency of usage. Main holts were in use on at least 60 per cent of his visits during the 18 month study, subsidiary holts on fewer than 30 per cent of occasions. At Lunna the main holts were significantly further away from the shore (median distance 90 m) than subsidiaries (median distance 30 m), and we did not have an obvious explanation for the difference. There was a striking difference in usage by different social categories of otters: independent subadult otters and non-resident, subordinate otters were never seen using main holts, always staying in inferior places (see Chapter 3).

Dispersion of otters along Shetland coasts in relation to resources

The main factors which are likely to affect otter distribution along the Shetland coasts are food, shelter, fresh water, and predators, apart from disasters and adverse conditions created by man.

The diet of Shetland otters (see Chapter 4; Herfst 1984; Kruuk and Moorhouse 1990) varies with season; in summer it is dominated by species of fish which occur mostly in sheltered areas, the voes (bays), species such as the eelpout *Zoarces viviparus* (see Chapter 5; Kruuk *et al.* 1988) In winter, however, rocklings of various species, as well as other, larger fish dominate, and they occur especially along more exposed coasts. It is clear from the map of home ranges of resident otters (Fig. 2.6) that each range includes both these main coastal types. Of course, this division is very basic, and both exposed and sheltered coasts have many different substrates and vegetation types, each likely to have different fish communities (Kruuk *et al.* 1988), to be optimally exploited by otters under different conditions.

If, indeed, the presence of different types of foraging areas is an important requirement for an otter range, then the 'grain size' of the landscape, that is the lengths of exposed coasts and the size of the bays, would become an important factor in determining the total lengths of home ranges. Otters can fish in waters down to over 10 m deep, quite a distance offshore, and in such a strip of water, with the fish densities as found in the Lunna area (Kruuk *et al.* 1988; see also Chapter 5), there is enough prey for more than one adult otter for every kilometre of shore. Thus, the overall 'carrying capacity' is quite high. It is likely that the only way to divide a stretch of coast into a number of ranges which each include both exposed and sheltered areas, and in which the number of otters is reasonably close to carrying capacity, is a system under which several otters share a range, that is a group range system.

The presence of sites, including holts, where otters can land and rest, is also likely to affect dispersion. However, the study area at Lunna was a coast with an abundance of sites to be used by otters when they were not at sea. There were many holts, used and unused, many lying-up sites between rocks, and from just watching the animals along the shores and on land it was unlikely that these aspects of habitat utilization posed a limitation on otter numbers and distribution there. The same was probably true for freshwater sites, places where the animals need to wash (see p. 198). It has been suggested that the size of freshwater pools, often simple small, deep holes in the peat, could cause them to be easily dirtied by a single otter having a bath. This, in theory, could cause competition between animals, if one otter were to leave a pool less suitable for use by the next (Kruuk and Balharry 1990; Kruuk and Moorhouse 1991). It could therefore affect range use and dispersion if there were only a few pools available. It is uncertain, however, whether this was actually the case anywhere.

Predation by wild animals on otters is rare, certainly in Europe where potential enemies (for example the wolf *Canus lupus*, lynx *Lynx lynx,* and various eagles) are few. The marine habitat appears to be somewhat more dangerous than inland areas in Britain, with potential predators including seals (for example the grey seal *Halichoerus grypus*), killer whales *Orcinus orca,* and white-tailed eagles *Haliaetus albicilla.* But man and his dog will always be the main enemy (see also p. 217). Seals are very common in Shetland (details in Anderson 1981), and we saw grey seals chase otters on five occasions, always in such a way that there seemed little doubt that the seals were after the otters' blood. In contrast, in the 10 interactions we saw between otters and common seals (*Phoca vitulina*) three times this seemed to be mere 'play' on the part of both participants, and on four occasions otters were chased really hard. But there was no indication that otters avoided the main concentrations of seals along the Lunna peninsula, that is the four haul-outs of seals, and they often fished or walked very close. Killer whales and white-tailed eagles were seen only rarely. There was no suggestion, therefore, of natural predators affecting the dispersion of otters.

The effects of people were more complicated. The otters were shy, and they avoided a person along the shore at distances of several hundred metres, as we knew to our cost. On the other hand otters regularly fished and swam close to places where people occurred 'normally', such as near fish-farms, near houses or farms, or near the bustling Shetland oil terminal, Sullom Voe. In fact in the oil terminal otters frequently stayed amongst the boulders below the huge loading piers, some even used the pump houses. Elsewhere otters may get disturbed by farmers or their dogs, or fish-farmers in their boats, but all our evidence from observations of radiotagged otters suggested that such disturbance was never more than temporary, and highly unlikely to affect the animals' range utilization in any substantial way. There had been no persecution of otters for many years, despite the many remains still found along the Shetland shores of the traditional otter traps, or 'otter houses' (p. 255).

Habitat and dispersion along sea coasts elsewhere, and conclusions

The association between otters and peaty coasts and fresh water which we found in Shetland, was useful also for understanding what happens in other areas. In Scotland otters occur all along the west coast, from Galloway northwards, along the shores of the Western Isles, Orkney, and Shetland, but they are virtually absent from the eastern seaboards (Green and Green 1980, and personal observations). Along the Orkney shore there are few otters, in sharp contrast to the abundance of the species in Shetland. At first we tried to explain this by possible differences in the abundance of prey. We were struck by the similarities between the distribution in Britain of black guillemots or tysties (*Cepphus grylle*) and otters, and tysties ate many of the same species of fish, such as butterfish and eelpout (Ewins 1986). This explanation for otter distribution might be superficially attractive, but it did not stand up to detailed scrutiny; for instance there was much potential prey, such as butterfish, rocklings, and various other fish, along the rocky shores of the Scottish east coast and in Orkney, but no or few otters. In Shetland, too, there were places where otters were few but where absence of fish was not likely to be the explanation.

There was, however, an association between the presence of otters and the absence of agriculture. Scotland's east coast, the north-east, and Orkney, all these areas with few otters were dominated by farming, whereas much of the west coast had moorland or woods going down close to the shoreline. That fitted with the Shetland experience, where along agricultural coasts such as the southern part of the Mainland of Shetland there were very few otters.

The Scottish west coasts, with their many otters, also have a much higher rainfall than the east, and the differences in agriculture are probably related to rainfall because of the wet leached soils of the west. Small freshwater pools are common along the Scottish western shores, with many showing intensive use by otters (Elmhirst 1938). Whether rainfall or lack of agriculture makes the west so favourable for *Lutra* is difficult to determine at this stage, it will need more detailed analysis. But other comparisons are more clearcut, between Shetland and Orkney for instance. There is almost no difference in rainfall (Berry and Johnston 1980), but Orkney is a region dominated by rich, well-drained farmland whilst agriculture in Shetland is only patchy. Taking into account the differences in the otter populations between the various areas in Shetland itself, I have no doubt that the comparative lack of otters in Orkney is related partly to human land-use and partly to a difference in geology which together result in very little surface fresh water along the sea shores in Orkney. Most of the Orkney coast is well-drained: misery for an otter in search of fresh water to rinse off the salt of the sea. Similarly along the coast of Skye, otters are closely associated with geological features (especially Torridonian sandstone) which determine the presence of freshwater pools (P. Yoxon, personal communication). Fresh water, in small pools or streams, appears to be the key habitat feature for otters along all sea coasts and I will discuss this further in Chapter 7.

Observations on spatial organization of otters in other areas were made along the shores of Mull, on the Scottish west coast, incidental to research on the energetics of otter foraging (Watt 1991). That study suggested a social organization similar to that in Shetland: several females sharing relatively small ranges in sheltered waters, with males ranging far more widely and overlapping with the female ranges. Along the coast of the Ardnish peninsula, close to the Isle of Skye, Kruuk and Hewson (1978) found a regular spacing of groups of holts, also suggesting some territorial system.

Organization has also been investigated in several other otter species which inhabit the sea-shore. The sea otter *Enhydra lutra* is not really comparable, being much more aquatic even than seals (Kenyon 1969). However, along the American Pacific seaboard the river otter *Lutra canadensis* lives in circumstances very similar to those of coastal *L. lutra* in Europe, although it is also predominantly a freshwater species (Melquist and Dronkert 1987). Large overlap in the coastal ranges of females has been reported for that species (Larsen 1983; Woollington 1984; Noll 1988), but precise boundaries and core areas were not studied. The female social system could be similar to that of *L. lutra*, therefore, but needs further investigation; the males of *L. canadensis* appear to behave somewhat differently. There are several observations of groups of males ('bachelor groups'), up to seven at a time, cruising along coasts in the Pacific (Larsen 1983; Melquist and Dronkert 1987), a phenomenon which has never been reported for *L. lutra*.

The marine otter *L. felina*, which occurs along the coasts of Chile and Peru, has ranges shared between several females, with the role of males uncertain (Ostfeld *et al.* 1989); details of the organization are yet to be studied. In the Cape clawless otter *Aonyx capensis*, a freshwater species which also occurs along sea coasts in South Africa, there is extensive overlap of coastal home ranges between individuals, both within and between sexes; female range boundaries were not known in detail, but several males shared, and they even foraged together (van der Zee 1982; Arden-Clarke 1986).

In summary, therefore, it appeared likely that the observations on otter ranges in Lunna, Shetland, were representative of otters in many coastal areas elsewhere, also of other otter species, although these other areas were not documented in as much detail. If the spatial organization of Shetland otters was an adaptation to resource dispersion, it was most likely that those resources would be prey availability and access to fresh water. The organization into group ranges with a patchy food dispersion is comparable to that in several other carnivores such as foxes (Macdonald 1981), badgers (Kruuk 1989*a*), and many other species (Macdonald 1983), but the system of core areas as demonstrated for the Shetland otters is unusual, and would not have been predicted from the Resource Dispersion Hypothesis (Macdonald 1983). This will be further discussed in Chapter 9.

ORGANIZATION IN FRESHWATER AREAS

Dispersion in fresh water: introduction

In the early, classical study of otters in fresh water Erlinge (1967, 1968) provided some excellent insights into the social life of the animals along rivers and small lakes in southern Scandinavia, based on snow tracking. The populations which he studied were quite dense, with on average one adult otter for every 4 to 6 km of river. The females lived in more or less exclusive areas, no more than 6 km of river, with some small overlap between neighbours. There were no group territories. Male territories were much larger, including the ranges of the females, and there was evidence of aggression between males on the borders of their ranges. Erlinge's otters were therefore a typical example of mustelid social organization (Powell 1979). By its nature his study was confined to winter, when the lakes were frozen over, and even streams were often covered in ice and snow. That may have been one reason for the relatively small range sizes he found; possibly, the otters ranged more widely in summer.

The first radio-tracking study on European otters was done in central Scotland by Green *et al.* (1984). Three individual animals occupied stretches of 16–22 km of river, but their actual home ranges could have been considerably larger still, as it was only possible to follow the animals for short periods. In that study there was evidence of other otters simultaneously using the same ranges as the focal animals. There were possible similarities with our Shetland results, therefore, and to analyse this further I will first discuss some of the habitat preferences of otters in rivers, streams, and lakes, then details of their spatial organization from our own studies.

Habitat selection in freshwater areas

In the north-east of Scotand otters are common in rivers, streams, and lakes, and we studied them especially in the two main river systems west of Aberdeen, the valleys of the Dee and the Don (Kruuk *et al.* 1993; Durbin 1993). The main objective of the research was to establish a possible relationship between otter and fish densities, and Leon Durbin carried out his PhD study on habitat utilization by otters. However, we also obtained information on spatial organization.

The rivers Dee and Don are both over 100 km long, and the lower sections, which we studied, were somewhere between 15 m and 100 m wide; the width of the tributaries could be anything between 0.5 and 5 m. Otters were and are found in almost every stream or lake in the north-east of Scotland, from the smallest upwards, even in the rivers right in the centre of large towns, such as Aberdeen. However, obviously they do have areas which they prefer or avoid.

Several authors have commented on the association between otters and woody vegetation on the banks (for example Jenkins and Burrows 1980; Mason and Macdonald 1986), but we could not substantiate this; the animals which we followed with radio-tracking showed no preference for woody bank vegetation, or for any other type of bank (Durbin 1993). The differences in results are probably due to methodology; previous authors used the distribution of spraints along rivers as indicators of usage of the streams, but we found spraint distribution unreliable for this purpose (Kruuk *et al.* 1986; Kruuk and Conroy 1987). Otters spraint near trees as dogs may use lamp-posts, and it is difficult to derive a habitat preference from that. The different results have clear implications for conservation management of otter habitat along banks (see Chapter 10).

An important habitat factor for otters in our study areas was the width of the stream or river, which was important in a way which was surprising and somewhat counter-intuitive. In the north-east Scotland river study we followed eight individuals almost daily, some for over a year, and we noted how much time the animals spent in tributaries, sections of tributaries, or sections of the main stem of the river of different mean width, so for each individual we estimated how it divided its time. Simultaneously, we could recognize the spraints of those individual otters, because we had given the animals a harmless isotope (zinc-65) which labelled their spraints for up to 6 months. By collecting all spraints from the tributaries and river sections we could estimate, therefore, for what proportion of total otter activity in those areas our radio-otters were responsible. From that, combined with the radio-tracking data, we could derive the total amount of otter time spent there, and compare it with the size of the stream, that is its mean width (Kruuk *et al.* 1993).

The results are shown in Fig. 2.14 and 2.15. First, we compared time spent per length of river and stream of different widths; it appeared that the otter population spent more time per length of river section along wider streams. As we had expected, however, there was quite a large variation in the results.

Nevertheless, it is somewhat unsatisfactory to look at the otter habitat in that way; there is obvious logic in comparing rivers and streams on the basis of area of water rather than mere length. We therefore expressed the amount of time otters spent per unit area of water along streams of different widths. That showed a quite different

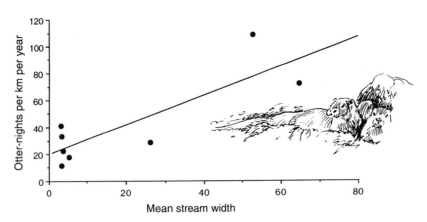

FIG. 2.14 Utilization of freshwater streams by otters in north-east Scotland, expressed as time spent per km of streams of different mean widths. Each point represents a mean value for a tributary or section of main river: $r = 0.83$, $r^2 = 0.69$, $p < 0.02$; $y = 18.5 + 1.08x$. Rank order correlation $r_s = 0.36$, not significant. (From Kruuk *et al.* 1993.)

result: an exponential decrease in otter use per hectare of water, with the size of the river (Fig. 2.15). Clearly the tiny, narrow streams are hugely important to the otters (Fig. 2.16). This is true even for those small waters running through intensively used farmland, or near

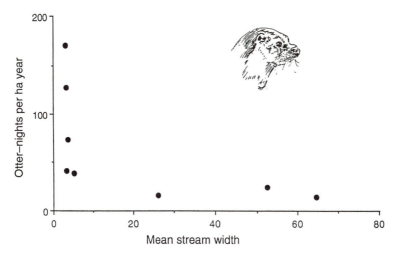

FIG. 2.15 Utilization of freshwater streams by otters in north-east Scotland, expressed as time spent per *hectare* of water in streams of different mean widths. Note difference from previous figure. Log/log: $r = 0.86$, $r^2 = 0.73$, $p < 0.01$; $y = 133.65x^{-0.59}$. (From Kruuk *et al.* 1993.)

FIG. 2.16 Small stream in an agricultural area in north-east Scotland: excellent otter habitat.

villages. The reason for this probably lies in the differences in fish density (Fig. 5.10, p. 146) and in the otters' habit of foraging close to, or under, banks (Durbin 1993).

In the streams and rivers, Durbin established that otters have clear preferences for different substrates: sections with riffles, large boulders, and/or with gravel are preferred over areas with sandy or muddy bottoms, or with stones. Again, the explanation appears to lie in the foraging behaviour, and it may well be related to the fact that our streams were inhabited mostly by salmonid fish. Carss *et al.* (1990) found that otters catch large salmon especially on riffles rather than in deeper waters (see p. 161).

Otters in Scottish freshwater lochs showed a different pattern in their use of banks and waters. Unlike their behaviour in running water, they discriminated strongly in favour of particular types of bank vegetation, especially reed beds. In some places there are wide fringes of reeds

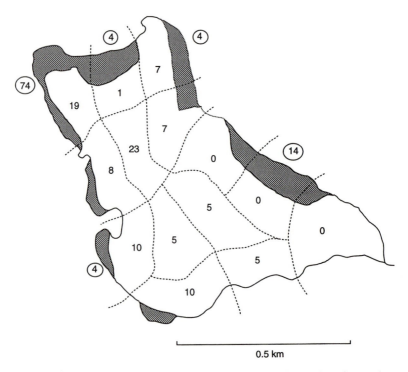

0.5 km

FIG. 2.17 Map of Loch Davan, Deeside, north-east Scotland, showing observation sections, percentage use of edge sections by four radio-tagged otters for day-time resting (circled, 131 observations), and percentage of prey captures observed from the same otters in the aquatic sections (99 observations).

(*Phragmites communis*) along the lochs, and otters spent most of their time in those when they were not fishing. As an example, we ranked the use of different equal-sized sections of shore of Loch Davan (Dinnet, Aberdeenshire) (see Fig. 1.6) by four otters with radio transmitters (Fig. 2.17). We then compared those section scores with (a) a ranking of the numbers of otter paths and couches, (b) the size of the reed bed in the sections, (c) the 'fishing distance index', the FDI, and (d) the distance from the nearest source of human disturbance, such as houses or main roads (Table 2.1). The FDI of a section was the sum of the estimated distances from sections where the otters went to forage, each distance multiplied by the proportion of all fish seen to be caught by otters in those foraging sites. Thus the smaller the FDI for a particular place, the closer was the point to sites with the highest fish capture value.

The results showed that the radio-otters used the area exactly as we could have expected from the distribution of otter paths and couches. The most important variable was the size of the reed bed—distance from feeding areas was not significantly correlated with usage. These

TABLE 2.1 Use of sections of shore of Loch Davan (Dinnet, Aberdeenshire) by otters, in relation to size of reed beds and distances from feeding areas and sources of human disturbance.

Section	(a) % Days spent (radio otters) $n=131$	(b) Ranking couches and tunnels	(c) Reed bed size (ha)	(d) Feeding distance index FDI	(e) Disturbance distance (m)
D 1	65	11	0.62	345	130
D 2	14	9	0.49	291	100
D 3	2	7	0.27	426	120
D 4	0	2.5	0	261	240
D 5	1	8	0.15	328	370
D 6	15	10	1.01	480	460
D 7	0	2.5	0	690	620
D 8	0	2.5	0	484	670
D 9	1	2.5	0.11	451	600
D 10	1	5	0.18	320	350
D 11	2	6	0.01	233	330

Spearman rank correlations: (a)/(b): $r = 0.91, p<0.001$; (a)/(c): $r = 0.91, p<0.001$; (a)/(d): $r = -0.20$, n.s.; (a)/(e): $r = -0.57$, n.s. (b)/(c): $r = 0.89, p<0.001$; (b)/(d): $r = -0.19$, n.s.; (b)/(e): $r = 0.55$, n.s.

otters, incidentally, always slept in couches or on vegetation when they were at the lochs (which was most of their time), never in holts, even in the middle of winter. Areas where the otters foraged, however, were the places with many eels, especially those with large eels (see p. 149).

Spatial organization in fresh water

Adult male otters spent most of their time on the main stem of the rivers, with frequent excursions up the tributaries; for instance, in the Dee and Don valleys, of eight large adult males which we trapped or radio-tracked, five were usually along the main stem of the river (Fig. 2.18), whereas of the 15 adult females and/or independent subadult otters, 13 were mostly along the tributaries or in lakes. As along the coasts of Shetland, the difference in habitat, of males using the main stem and females the tributaries, was by no means absolute, only a statistically significant preference.

Within each home range we frequently saw or heard encounters between the animal which we were radio-tracking and other otters, anywhere in their range, and not just near boundaries. In the case of the adult males there was good evidence that they were seeing several females in the different tributaries and on the lochs. However, we had no good evidence that home ranges of females on the tributaries overlapped; there was only the fact that there were several other otters of unknown sex and status using the same area as the target otter in our

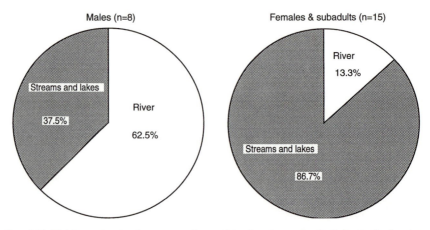

Fig. 2.18 Habitat selection by otter males, and by females and subadults, in freshwater areas in north-east Scotland: sites where different individuals were captured in box-traps. The difference is statistically significant: Fisher exact probability test: $p < 0.05$.

radio-tracking studies (Kruuk *et al.* 1993). This was similar to what Erlinge (1967) had found in Sweden. On the lakes, such as the Dinnet Lochs in the Dee valley, there were always several females using exactly the same waters at the same time, sometimes together and without overt aggression, even when they were foraging with litters of cubs (as observed also by Jenkins (1980)); several times I saw six or more otters simultaneously.

The size of otter home ranges in freshwater habitats is quite staggering, especially of the males—but this is only because we express this size in an unusual way, as length of water course. Green *et al.* (1984) found the length of range for an otter male to be 40 km, and one of Leon Durbin's males moved around in 84 km of stream. The mean length of river and stream used by otters in our north-east of Scotland study was 34.8 km for males (*n* = 6) and 20.0 km for females (*n* = 2), close to the results of Green *et al.* (1984) of 40 km for a male and 18 km for females (*n* = 2). However, in terms of area of water the figures are somewhat more like those we would expect for animals of this size: the mean range of fully adult males was 63 ha of water, of the females 20 ha of water (Kruuk *et al.* 1993). The ranges of badgers in the same region are around 50 ha (Kruuk 1989*a*).

Holts and couches in freshwater areas

Whilst otters along the Shetland coasts appeared to be dependent on elaborate, specially constructed holts (p. 43), the freshwater animals were strikingly different. On loch shores otters almost never rested or slept underground, but always in the reeds or thick vegetation. They slept on 'couches' (Fig. 2.19) or just curled up somewhere, even in mid-winter, regardless of the frequent misery of the Scottish climate. Reeds appeared to be the favourite bank habitat for otters, but in Scotland also rhododendron, which occurs as a weed in many areas, is frequently used. Couches have been described in detail by Hewson (1969*a*). Measuring 0.3 to 1 m across, couches are made by otters biting or pulling off vegetation, and then dragging it over some distance before flattening it to curl up in. Vegetation is carried in the mouth, or sometimes held with the forepaws and dragged backwards, like badgers (Neal 1986; Kruuk 1989*a*), and the main purpose of the couch appears to be to provide a dry base. As a considerable refinement, we found that on the Dinnet lochs breeding females made special covered couches, consisting of a large ball of reeds (Fig. 2.20). These covered couches were somewhere on a dry spot in a large reed bed but often close to water, with one or two entrances at the sides, and a lining of soft grass—comparable to

FIG. 2.19 Normal otter resting couch in a reed bed along a Scottish loch.

FIG. 2.20 Covered 'natal couch' in a reed bed. Side entrance; the roof caved in after the otter family abandoned it.

nests of birds such as magpies (Taylor and Kruuk 1990) and first described by Buxton in 1946. One of our radio-otters gave birth to two cubs in such a natal couch, which was about 1 m in diameter, 0.5 m high and about 2 km away from the nearest open water where the other otters were living.

Along the rivers and streams otters were more inclined to sleep in holes, proper dens, than they were at the lochs. Even so, these holts were less important in that habitat than on the sea coasts; if there were small islands in the river, otters appeared to prefer sleeping above ground on those islands rather than in a bank-side hole. It also struck us that when otters did sleep underground, this was often in some man-made construction, a heap of boulders, a culvert, a drain pipe, or an artificial road embankment. Upturned roots of windfall trees were popular daytime sleeping sites, and one male otter frequently slept in the wreck of an old Vauxhall car which was half submerged along a very small stream.

Proper holts were few and far between along the rivers, probably largely because otters had little use for them; the Shetland otters had shown us that they have no problem in digging quite large holts themselves. But some individuals were more likely to sleep underground than others, as a purely individual preference. For instance, observations on two adult males which we had followed along the River Dee in 1990 showed that one spent 23 per cent of 84 days in holts, and the other 37 per cent of 214 nights. This was a significant difference ($\chi^2 = 4.96$, $p < 0.05$). The observations on both otters were spread over different seasons, and both were more likely to be in holts during the winter (24 and 45 per cent) than the summer (17 and 8 per cent).

Bank-side holts were relatively simple affairs compared with the Shetland holts; in fresh water they almost always had just a single en-trance, within a few metres of the water, and more often than not we found them when radio-tracking in places which otherwise we would have overlooked. Holts could be in open, grassy banks or in woodland or scrub, there was little obvious selection of habitat. We never found a holt with an underwater entrance which has so often been described as 'typical' in the literature. Interestingly, the use of holts by *L. canadensis* is very similar to that of *L. lutra*; they were not important in freshwater habitats and were never dug by the otters themselves (Melquist and Hornocker 1983), in contrast to holts of *L. canadensis* along the Pacific coast (Larsen 1983; Woollington 1984; Noll 1988).

SPATIAL ORGANIZATION: SOME GENERALIZATIONS AND COMPARISONS

Given certain habitat preferences, for example for cover when resting or for narrow streams when foraging, the spatial organization of otters in fresh water appeared to fit into the same general pattern as that along the sea coast. The animals show a pattern of home ranges with an intrasexual spacing-out mechanism and, at least under some conditions (on lochs for instance), a total overlap of female ranges. There may be a similar system of core areas in fresh water as in the sea, but this will need further study. There also appeared to be differences in spatial organization: for instance, females with cubs in freshwater lochs were more tolerant towards each other than those in Shetland, and on several occasions I saw families foraging together without any animosity or avoidance, which was rare in Shetland (Kruuk and Moorhouse 1991; see also p. 81).

It is clear that otter densities in different habitats should not be compared simply as numbers of otters per stretch of bank or coast, which has led to the suggestion that there are much higher otter densities in marine habitats (Kruuk and Hewson 1978; Mason and Macdonald 1986). Coastal otters use a much wider strip of water, going as far as 100 m off shore, and numbers are estimated at up to one animal per kilometre (Kruuk *et al.* 1989), whilst inland animals use streams often no more than 1 m across, and average numbers of one per 15 km are suggested (Kruuk *et al.* 1993). These figures mean little in terms of otter density, but on the other hand it is difficult to calculate the number of otters per area of water. This is true especially along the coast, where otters fish mostly close to the shore, and progressively less at greater depths. There is no clear-cut solution to this problem, but I will return to it in Chapter 8.

The organization of several other species of otter appears to be rather similar to that of *L. lutra*. In a study of *L. canadensis* in Idaho, for instance, home ranges in rivers were generally similar in size to those in Scotland, often two females used the same range, frequently individuals or family parties were seen together for some time, especially when fish were plentiful, and rarely was there any aggression (Melquist and Hornocker 1983). Resident males used up to 63 km of river, and also it was suggested that females sharing a home range had different 'activity centres', clearly comparable to the 'core areas' of our Shetland animals. The main difference in social organization between *L. canadensis* and *L. lutra* appears to be the occurrence of male groups, 'bachelors', in the former species where it lives in the sea (see p. 51).

In Thailand I observed three species of otter sharing the waters of the river Huai Kha Khaeng: *L. lutra*, *L. perspicillata* (smooth), and *Aonyx cinerea* (small-clawed), in a system of intra- and interspecifically overlapping home ranges, not substantially different from the single-species organization (Kruuk *et al.* 1994). But some of the other species of otter show a substantial departure from this, for instance the South American giant otter *Pteronura brasiliensis*, which lives in packs of up to 20 animals (Duplaix 1980). Their troops in the forest rivers are extremely noisy, extended families consisting of several adult males and females as well as young ones. Recent studies suggest a usual pack size of fewer than nine individuals, made up of an adult male and female, and offspring from the two previous years (C. Schenck and E. Staib personal communication). It seems likely that in this species the brief association between males and females which we found in *L. lutra* has evolved into a permanent one, and that male and female ranges have become identical in a process similar to that found in the European badger (Kruuk 1989*a*). That animal, also a mustelid, shows an organization of groups ('clans') with several males and females, which has evolved from a solitary social system with independent and larger male ranges.

Alternatively, it is possible that the giant otter social system is comparable to that of the brown hyaena *Hyaena brunnea* (Mills 1990), where groups of related females and males defend a group territory, but where all the mating is done by itinerant males, a fundamental difference from the badger system but a relatively small evolutionary step for other social organizations. It would be fascinating to study exactly what role the distribution of resources plays in the social system of giant otters, what variation there is between different populations. Such questions about evolutionary differences within and between species are more than a merely academic exercise: it is important to find out what options each species has in responding to the availability of resources and the way in which these resources are dispersed. Could our European otter perhaps also live in packs, given the right environment and the right availability and dispersion of fish?

There are other species of otter which are known to live in large packs, at least sometimes. The spotted-necked otter *L. maculicollis* in Africa is seen in groups of up to 20 (Procter 1963), although the largest number of animals I have seen together was 10 (Kruuk and Goudswaard 1990). All the observations on large groups come from the East African lakes, especially Lake Victoria, where the animal is diurnal; it appears to be much more solitary and nocturnal elsewhere, as in South Africa (Rowe-Rowe 1978). I suspect that this is related to the nature and dispersion of food, but it could be that something completely different is having an effect: packs could be a protection against predation, as in some species of mongoose (Rood 1986). Predation pressure on this small otter species could be heavier in the more open lake environment, where there are many eagles, crocodiles, and other predators. I noticed that within groups of spotted-necked otters, individuals often foraged solitarily, but when they covered longer distances along the lake shores they swam in tight packs (Kruuk and Goudswaard 1990). The spotted-necked otter would be an ideal species for the study of such problems of sociality, effects of predation, the role of cover, and the distribution of resources.

The best-known social system in any species of otter is that of the sea otter, *Enhydra lutris*, in the northern Pacific (Kenyon (1969) and others; review in Riedman and Estes (1990)). It is also the most aberrant species: much larger (up to 30 kg for males, about three times the weight of a European otter), and carrying out virtually all of its activities in the water. Its social organization is totally different from that of any other mustelid, and is reminiscent of some of the seals. The animals may be seen resting, floating on the surface, in groups that occasionally number several hundreds (once of about 2000) (Estes 1980). These groups are called 'rafts' and consist of only males, or of only females with or without dependent young, and 80 per cent of observations of sea otters resting were of animals in groups. They forage on their own, however (98 per cent of observations), away from the rafts, and also parturition and mating are activities carried out away from others (Riedman and Estes 1990). At least some of the males maintain territories (Garshelis *et al.* 1984); presumably, the large groups of males are 'bachelors', not reproductively active whilst living in rafts. The large groupings are not just random aggregations, despite the fact that individuals and small groups come and go from the rafts; there are observations of groups of individuals 'helping' each other, for example in escaping from nets or predators (Riedman and Estes 1990).

The organization of this species is not only quite different from that of any other otter, it is also highly variable along the Pacific coast. Perhaps the difference between sea otters and the others can be understood in terms of dispersion of their resources, or perhaps it is related to differences in predation, although this has not been studied yet. Sea otters are capable of diving much deeper than the other otters (down to 100 m) (Newby 1975), and they use large stretches of open water rather than merely the coastal strip, feeding mostly on invertebrate prey (but also on fish) (Estes *et al.* 1981). Obviously food distribution will be very different from that of a *Lutra* species feeding closely inshore, and the sea otters also are not dependent on other resources such as fresh water

(p. 196), and holts. The intraspecific variation of sea otter organization, related to the dispersion of resources, should give some fascinating clues about the evolution of their communities.

SPATIAL ORGANIZATION AND RESOURCES

The relationship between the nature and dispersion of resources (food, water, shelter) on the one hand, and that of the otters on the other, will be discussed in more detail in Chapter 9 after both the resource distribution and the mechanisms of resource exploitation have been analysed in further detail. However, some generalizations may be useful here. For instance, we found that in the sea as well as in fresh water, male otters use a habitat which is different from that of the females: more exposed coasts, deeper water, larger rivers. This is not just a more 'macho' performance in areas where prey sizes are larger because as will be shown later the food resources in these male areas are less productive. A comparable sex difference in habitat use has been observed in red deer (*Cervus elaphus*) Clutton-Brock 1982): the larger and stronger sex is found in the least favourable areas. As a generalization, females divide the habitat between themselves, males fit in somehow.

The most obvious generalization of the social system in *L. lutra* is that it is variable, with females defending individual or group ranges which are highly different in size and habitat (from tiny streams to open sea coasts), and individual otters of either sex ranging between 4 and 80 km distance.

Such variability, with optional group territories, has also been found in some other carnivores, several of them sharing the same parts of the country with the otter, such as the red fox (Macdonald 1981) and the European badger (Kruuk 1989a). In these species the variability in dispersion has been related to the distribution of food resources (Macdonald 1983). It has been argued that animals live in groups because their resources occur in unstable patches, for example in sites of high food density; several such sites are needed within one territory in order to have access to resources at all times. In such a territory with several resource patches a number of same-sex individuals can be tolerated without substantial competition. Such a relationship makes intuitive sense, especially where otters have been shown to exploit food 'patches' (Kruuk *et al.* 1990), but there are problems with this explanation, such as the importance of phylogeny, and the detailed division of home ranges into core areas, which will be discussed in Chapter 9.

As a generalization, however, the Eurasian otter has a flexible organization, a range of social systems which includes the basic 'mustelid organization' (Powell 1979) of simple, individual intrasexual territories, with male territories larger than those of females, and with the additional option of group territories and individual core areas within those territories. Other species of otter, with the exception of the sea otter, have organizations which also fit this description, but with differences between species, some being more gregarious than others. Whether such differences are species-specific or resource-specific is as yet an important and open question.

3

Social behaviour

Despite a social system of group ranges otters are strikingly non-social. 'Sprainting' (scent marking) probably enables them to avoid the others who are living in the same range in order to exploit the fish and other resources without interference. Sprainting appears to function as a means of communication mainly over periods lasting only a few hours; it is associated with feeding and other forms of resource utilization, and all categories of otters do it almost equally often, but much more in winter (when resources are low) than in summer. If otters meet, there is little aggression (except between territorial males), but more avoidance. Sexual behaviour is described; otters are polygamous and polyandrous. The two-stage strategy of maternal care for the cubs (first in a natal holt, then accompanying the mother) is described; some cubs are abandoned. The male is not involved in rearing the cubs. Most social interactions are seen within the small families of a mother with her cubs. Mothers appear to teach their offspring to fish, and cubs get the larger prey whilst the mothers eat small fish. Cubs are dependent on their mother for 10–16 months, an unusually long period which may be related to the acquisition of fishing skills.

INTRODUCTION

One of Niko Tinbergen's important contributions to science (Tinbergen 1951) was his insistence on the analysis of behaviour patterns in terms of what he saw as the three main frameworks behind questions in biology: that is, causation, effects ('function'), and evolution. Spatial organization can be considered in a similar way, and here I will be concerned with its causal aspects, the mechanisms and behaviour patterns important in the maintenance of the organization.

In the previous chapter otter group ranges were described, with several families within a range, in Shetland each female or family having its own core area. It is a complicated spatial system, probably somewhat different from that in freshwater areas, and the result of interactions between individuals over the years. What are the behavioural strategies which are used to maintain this social system, how do the otters react to each other, what actually happens in the field? Is otter behaviour optimally adapted to maintenance of the spatial system, and to the pattern of exploitation of resources?

It was a fairly rare occurrence for us to see otters face to face with each other (except for mothers and their offspring), despite the fact that several animals used the same stretch of coast. The otters appeared to be very adept at avoiding confrontation, a phenomenon in which the otters' system of communication must play a dominant role. Nevertheless, over the years in Shetland we saw almost a hundred interactions apart from those within family groups, and they gave us at least some idea of how complicated and subtle the relationships are. Unfortunately it was less often that we knew the previous history and family ties of the participants in a particular interaction, an essential prerequisite for a full understanding of the social system. The following is an example of an observation of such an interaction.

February 1986, mid morning on a quiet, rather dull day, and I am watching a small island, about 80 metres from the shore in the range of the Lunna Voe otters. One otter lies asleep on the seaweed, a couple of metres from the water's edge. Along comes another, swimming close to the shore. They are the same size, both females, and they know each other well as they are mother and daughter: the sleeping one is Baboushka, mother of the new arrival Weibka, who is now entirely independent but lives in the same home range. Weibka comes to shore downwind of Baboushka and walks up to her; Baboushka lifts her head then curls up again. Weibka licks her mother's back, but then jumps away at Baboushka's response. Baboushka gets up, walks off along the edge of the water, with Weibka following her with a loud, continuous high-pitched tremulous call, 'wickering'. This call is usually heard in a somewhat defensive context; here, it suggested the ambivalence in the relation between the two animals, the attraction between close relatives whilst at the same time there is some spacing-out mechanism at work. Soon the two otters are both fishing again, each about ten metres offshore and drifting off in opposite directions; probably they would not meet again for several days.

The family background was important, and observations such as this one gave clues about how the spatial organization worked in practice. In the following pages I will describe and quantify some observations which are related to the otters' pattern of spacing, including aggression and territoriality, and their sexual and parental behaviour.

In fact, there are remarkably few visual displays and calls, just as was found in other, more social, mustelids such as sea otters (Kenyon 1969), Eurasian badgers (Neal 1986; Kruuk 1989a), or all the more solitary ones (Sandell 1989). The most articulate and most vocal mustelid is the South American giant otter (Duplaix 1980). In general the Carnivora, and especially also the Mustelidae, make little use of displays (in the widest sense of the word) compared with, for instance, ungulates (for example Estes 1967, 1969; Jarman 1974; Clutton-Brock et al. 1982) or birds (for example Tinbergen 1960). Possibly, the disadvantage of displays being conspicuous are greater in Carnivora; possibly also, the nature of their food resources and the way in which these are exploited enable carnivores to get by without elaborate displays (see below). Nevertheless, there are species such as the spotted hyaena which are extremely noisy and have many obvious displays (Kruuk 1972).

In the spatial organization of otters olfactory communication probably plays a key role, its function being mostly to keep otters spaced out. This system of communication, sprainting, also provides the first, best-known, and often the only sign to naturalists in the field that there are otters around.

SPRAINTING BEHAVIOUR

An otter spraint is no more than a small dollop of faeces, when fresh usually black and tarry with fish bones, often less than 1 cm across. The insignificant, fishy-reeking object is incontrovertible evidence of the passage of an otter, and all that most people will ever see of the elusive shy nocturnal animal. Otters tend to spraint on vantage points or other striking places, for instance on the top of prominent rocks along the water's edge (Fig. 1.15, p. 22), under bridges and near trees, or at the junctions of tributaries. Along sea coasts many spraints are found near small pools (Fig. 7.6, p. 199), near holts (Fig. 3.1), or on promontories. The relevant observation is that they are often conspicuous, and otters really go well out of their way to deposit a spraint on just such a prominent point. A spraint is so small that an otter will have to produce many of them each day to satisfy the function of elimination of food remains. Together, these observations imply that spraints have another purpose, probably scent communication (Erlinge 1968; Kruuk and Hewson 1978; Mason and Macdonald 1986; Kruuk 1992). Spraint sites are often used for many years and, because of the high nitrogen concentrations, many attract a characteristic flora of green algae along the coast, and of nitrophilous grasses such as Yorkshire fog (Holcus lanatus) inland, the dark green colour of the grass on and around the site standing out conspicuously.

FIG. 3.1 Spraint with crab remains, near holt entrance.

Spraints are interesting for two totally different reasons. First, as in many other carnivores, scent marking with scats is likely to feature prominently in the maintaining of territories. A great deal has been written about this (see the summaries in Macdonald 1980a, 1985; Gosling 1982; Gorman and Trowbridge 1989), and for species such as spotted hyaenas and badgers much was learned about their territorial systems just from studying the distribution of faecal scent marks (Kruuk 1972, 1989a; Gorman and Trowbridge 1989). The striking and intriguing organization of otters into group territories and individual core areas suggests that in this species, too, knowledge of scent marking could provide insight into the mechanism of the dispersion pattern.

The second reason for our interest in spraints is a more pragmatic one: many research workers have used them to monitor otter populations, to establish trends in numbers and habitat preferences (for instance Lenton *et al.* 1980; Green and Green 1980, 1987; summary in Mason and Macdonald 1986). However, it has not yet been established that spraints can be used in this way, and such surveys are based on an untested assumption, that is that if there are more spraints along one section of coast or river in comparison with another, then otters will have spent more time there. If we could verify this, spraint surveys might be a powerful tool with which to assess otter numbers and activity; their use, however, has been called into question (Kruuk *et al.* 1986; Kruuk and Conroy 1987; see also Chapter 8).

For these two main reasons I spent some time in Shetland investigating sprainting behaviour (Kruuk 1992). Why do otters spraint, who does it, when, and where? If it were important in sexual behaviour, one might expect differences in sprainting between males and females, and it would also be seasonal where reproduction is seasonal. If it played a role in territorial defence, boundary areas might show concentrations of spraints (Gorman and Mills 1984; Gorman and Trowbridge 1989).

Some relevant work had already been done before we started in the Lunna study area, especially by Jim Conroy. He found that the number of spraints which he collected along stretches of coast of the Yell Sound in Shetland varied widely between his bimonthly visits, and the use of sprainting sites along different sections of coast often fluctuated quite independently. There was also a huge seasonal variation in spraint numbers, and about 10 times more spraints were found per visit in winter than in summer (Conroy and French 1987). The same seasonality had also been found elsewhere, in English rivers (Macdonald and Mason 1987). Reproduction in Shetland otters is highly seasonal (Kruuk *et al.* 1987; see also Chapter 8), but in English fresh water it is not, with otters giving birth at any time of the year (Harris 1968). If sprainting had anything to do with reproduction, then the Shetland spraint seasonality might have been expected, but not the fluctuations in numbers of spraints along English rivers.

In the Shetland study area, the seasonal differences in numbers of spraints along the shores were caused not by otters spending more or less time there during winter or summer, but by otters actually sprainting relatively more often on land in winter (Kruuk 1992). I calculated how often I saw an otter deposit a spraint (Fig. 3.2) for every hour that it was observed, and also for every time I saw an animal come ashore. About 20 different animals were involved. On both scores there was a striking seasonality, including a spectacular difference between the summer months and the months when winter merged into spring (Fig. 3.3). Sprainting frequency in March was 12 times higher than in June. This was not because otters produced fewer scats in summer, but during that time I actually saw them defaecate in the water much more than in winter (they lift their tail out of the water when they do so; aquatic defaecation is therefore often quite easy to observe). All categories of otter, males, solitary females, and females with cubs (Fig. 3.4), showed the same pattern, and there was no difference between them in the seasonal rates of sprainting on land (Fig. 3.5).

Sprainting was therefore unlikely to have a function related to sexual behaviour or reproduction. Was it some simple territorial behaviour? Most likely, it was not. Four territorial boundaries were accurately

FIG. 3.2 Sprainting—note tail position.

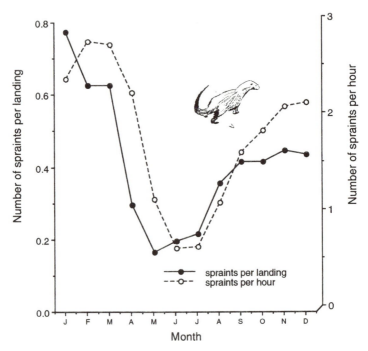

FIG. 3.3 Frequency of sprainting behaviour, at different times of year. Three-point moving averages. Total number of spraintings observed 292. Significance of seasonality: $\chi^2 = 117.3$, d.f. $= 9, p < 0.001$. (After Kruuk 1992.)

FIG. 3.4 Female with cubs sprainting on exposed sea weeds.

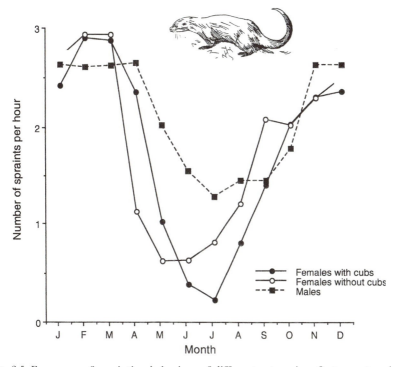

FIG. 3.5 Frequency of sprainting behaviour of different categories of otters, at various times of year. Three-point moving averages of numbers of spraintings per hour of observation. The differences between males and females were not significant (Wilcoxon test, $p > 0.5$). (After Kruuk 1992.)

known, and when the sprainting rates in the sections of coast at either side of these boundaries were counted, they proved to be no different from sprainting rates anywhere else. In the 350 m sections near boundaries, I saw 5.2 spraints deposited per section (± 4.8 standard deviations; number of sections $n = 8$), whereas overall along the coast this figure was 6.3 (± 6.3 s.d., $n = 46$; Kruuk 1992). Expressing it differently, near borders otters deposited an average of 0.26 spraints each time they landed, elsewhere 0.37. None of these differences were significant. This faecal scent marking of otters showed a pattern, therefore, which was at variance with what I had seen in badgers or spotted hyaenas, where latrines were strikingly crammed along the border-lines of the clans (Kruuk 1972, 1978, 1989a; Mills 1990).

Moreover, it has been argued convincingly by Gosling (1982) that scent marks are effective in territorial defence especially during aggressive encounters between intruders and territory owners. During an encounter, an intruder can identify an owner after matching the latter's scent to the smell within the territory. This would render escalation of such an aggressive encounter into a full-scale fight less likely, and it is suggested that the intruder desists after assessing the extent of the owner's 'investment' in the territory. This 'match hypothesis' appears very plausible, and explains many features of scent marking. However, in the case of otter sprainting it probably does not apply. There appears to be very little physical aggression in defence of the territory, very few one-to-one interactions, especially not in females, unlike many other carnivores such as badgers or foxes.

But there is an alternative explanation for the biological function of sprainting, independent of sex or territoriality. It seems plausible that these scent marks serve as signals within the otter group territories, between the group members or any other otters fishing along the same coast (Kruuk 1992). With spraints along the water line, otters may simply signal to any others that they are, or have been, feeding in a particular site or stretch of coast or stream. For such a mechanism to be effective, it is not necessary for otters to be able to recognize individuals from their spraints, although it is known from experiments in captivity that they can do so (Gorman and Trowbridge 1989).

Individual otters are animals of habit; they come back to the same feeding sites (see p. 168), or to the same small washing pools or holts (see p. 198), day after day. It will be advantageous for a second otter, arriving later, to go elsewhere because the site will have been used, or is in use, by an otter who has the advantage of prior knowledge. If this second otter does go somewhere else, the first otter will benefit because then it can forage without competition, or it can return to that site before long (often the next day) and should find the resource less depleted if no other otter has been there in the meantime. The same argument could apply to sprainting next to a small freshwater pool along the coast, or fresh water in a holt: if many otters were to wash in it, the water could become too salty or dirty. An otter advertises by sprainting that it has been bathing there, which might induce others to keep out, to mutual benefit.

The observation that all otters—males, females, and cubs—spraint at more or less the same rate, and that the animals would spraint anywhere in their range where their resources could be, is consistent with the idea that sprainting means advertising the use of a resource. Also consistent is the observation that many spraints, about 30 per cent, along the coast are deposited in the intertidal area, so they can only function for a few hours at the most. If this comparatively simple explanation, of spraints preventing competition, would hold true, one could expect that otters would spraint (a) near places where they were fishing, that is before, during, or after feeding bouts and (b) near other resources where competition between individuals could arise. A logical consequence of this hypothesis would be that (c) sprainting should occur especially at times when resources were scarce. Finally the explanation would hold especially if (d) otters exploited a system of patchy, replenishing resources in their home range, not just feeding randomly throughout, so they would derive some advantage from persuading other otters not to use, for instance, a food patch after they had fed there themselves.

To test these ideas, Andrew Moorhouse and I noted what otters were doing immediately before and after they sprainted (Fig. 3.6). There was no doubt that fishing or eating and sprainting were closely associated. In 66 per cent of the 331 times in Shetland that we saw otters sprainting, it was either immediately preceded or followed by a feeding sequence. An otter's feeding bout, that is a single foraging session in any one area, was always preceded by sprainting, but sprainting occurred significantly less often during or after such a feeding bout (Fig. 3.7). For resources other than food, the close association between spraints and holts or freshwater pools has been mentioned already.

The striking seasonality in sprainting behaviour in Shetland (Fig. 3.3) coincided with the annual fluctuation in food availability (Kruuk *et al.* 1988): there was a peak in numbers of potential prey in midsummer (when there are fewest spraints), and a trough in winter and spring (the times of most spraints). The same appears to be true in freshwater ecosystems, where productivity in fish populations is directly related to water temperature, and fish biomass is, therefore, highest at the end of

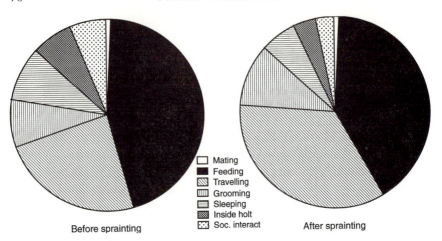

FIG. 3.6 Behaviour of otters before and after sprainting (percentage of 131 observations).
(After Kruuk 1992.)

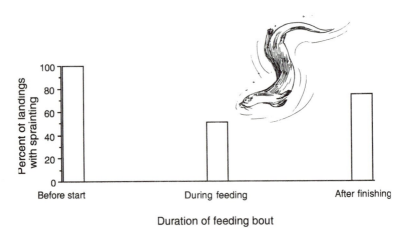

FIG. 3.7 Frequency of sprainting during landings immediately before, during or after a
feeding bout. Statistical significance: before and after, $\chi^2 = 17.9$, $p < 0.001$; before and
during, $\chi^2 = 46.6$, $p < 0.001$, during and after, $\chi^2 = 13.3$, $p < 0.001$. (After Kruuk 1992.)

the summer and lowest in spring (Kruuk *et al.* 1993; see also p. 145).
Spraint densities in such habitats fluctuated correspondingly (Erlinge
1967, 1968; Jenkins and Burrows 1980; Murphy and Fairley 1985; Mason
and Macdonald 1987).

For the hypothesis on the relation between spraints and resource utilization to hold, we still need to demonstrate that otters exploit their

resources in such a way that it would be beneficial for them to signal this use to others. The use of food patches will be discussed in detail in Chapter 6; this shows that otters use the same feeding sites again and again, and that coastal sites which were cleared of fish were repopulated with potential prey within 24 hours. From the way in which otters forage, I have no doubt that they know every nook and cranny in their regular range. It would make sense, therefore, for an otter to signal to others to keep out because it would be detrimental for them to arrive at a known food patch to find that someone else has just depleted it. Just as important, it would make sense for the receiver of the signal to go elsewhere, because chances are that another otter has just denuded the place of its resources.

With this system of signalling, otters possess a mechanism whereby they can partition resource utilization within a group territory without direct confrontation. It is a system that appears to be especially well adapted to the use of resources which a carnivore exploits with some prior knowledge and experience, and where for instance prey is caught again and again at the same sites. It may well be that many other species of carnivore use scent marking in a similar strategy.

As far as is known, all other species of otter use spraints for communication, probably in a similar way to *Lutra lutra*, except the sea otter *Enhydra* which spends very little time ashore (Kenyon 1969). Details of spraint sites may differ between species. The North American river otter spraints in much the same manner as *Lutra lutra*, both along sea coasts and along streams (Melquist and Hornocker 1983), but for the giant otter in South America, Duplaix (1980) described very large areas, of many square metres termed 'camps', almost covered with spraints. Similarly the sprainting sites for the smaller, but also group-living spotted-necked otter in Africa has large, very smelly sites with masses of faeces and urine scattered around (Kruuk and Goudswaard 1990). The Cape clawless otter has conspicuous spraint sites both in fresh water and along coasts, using them in a similar manner as coastal *Lutra* in Shetland, even to the extent of also sprainting at freshwater pools along the shores of the Indian Ocean (Arden-Clarke 1986; and personal observation).

I collected some interesting observations on sprainting in other species when in Thailand, where *Lutra lutra*, *L. perspicillata* (smooth), and *Aonyx cinerea* (small-clawed otter) occurred together, along the river Huay Kha Khaeng (the upper ranges of the 'River Kwai') (Kruuk *et al.* 1993*a,b*). Their main spraint sites were somewhat different from each other, as we discovered from tracks and direct observations, with the last two species sprainting usually high up on large, flat rocks (but not always) and the smooth otter in more prominent sites than the common and the small-clawed. The common otter sprainted lower down near the water, with fewer spraints per site than the others produced, but in the same type of place as this species does in Europe. The smooth otter produced large, extremely smelly sites with masses of faeces, and the small-clawed spraint sites were conspicuous because of the many crab remains in their faeces. Especially fascinating was that, although there were species differences in the kind of spraint sites they established, they did visit and sniff each other's, as we determined from tracks. They also had a clear overlap in diet, and if these spraints do have a function of preventing competition, then such interspecific interest in spraints is exactly what one would predict.

Otters also use other means of scent communication; for instance urination must be very important for signalling. The animals often urinate on spraint sites, but we know little about its significance. Quite likely it will play a role in communication between the sexes, at least sometimes; urine from females is known to convey information about oestrus in dogs (Beach and Gilmore 1949), and both cats and dogs can detect sex from urine (Verberne and de Boer 1976; Dunbar 1977). Otters also rub their cheeks against stones, they roll on special rolling-sites, males scrape up small heaps of sand or vegetation at, or close to, sprainting sites: all these behaviour patterns are probable means of scent communication, as in other carnivores (Rasa 1973; Gorman and Trowbridge 1989), but the exact messages which they convey are still quite unknown.

AGGRESSIVE BEHAVIOUR

A system of exclusive home ranges is likely to be based on some kind of aggressive interaction between the inhabitants. This has been demonstrated for simple territories of solitary animals (Sandell 1989), clan territories of spotted hyaenas or badgers (Kruuk 1972, 1989a), and I would expect it also for female group territories of otters. However, at least in Shetland otters, such aggressive interactions were remarkably rare, and despite many years of observations our information on aggressive behaviour is still rather poor.

Really hard fights I saw only between males, and conversely, whenever males met, they almost invariably fought (eight out of nine observations; Fig. 3.8). The following is an observation of such an interaction:

January 1986 in Shetland, in the fading light of a dull afternoon, about four o'clock. Following a large male which swims along the coast in the north-west part of our study area, I see him land on a small rocky island, just out of sight between some large boulders. Within minutes another otter arrives, from the same direction as the first—it must have been following, without me being aware of it. The other one is also a big animal, obviously male, swimming about five metres off-shore, with its tail showing conspicuously along the surface. He passes the place where the first one landed, clearly gets wind of the other's presence, suddenly switches direction, and lands. A fleeting sniff of each other's faces, then a loud, high-pitched screaming and 'wickering', probably from both animals. A lunge, followed by an incredibly fast chase, off the rocks, into water, on to land, up the slopes directly in front of me. Screams, the fleeing otter is overtaken, bitten hard in the rump. A turn-around, again a chase in that curious lolloping gait, though very fast. This all takes some seconds only, then

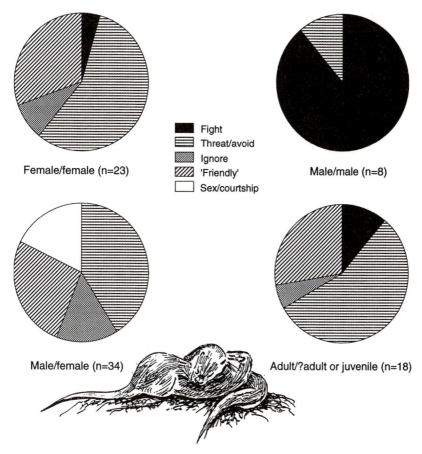

FIG. 3.8 Interactions between different categories of otters. Males were significantly more aggressive to each other than females (Fisher test $p < 0.05$). (Data from Kruuk and Moorhouse 1991.)

the totally pre-occupied animals run straight up to me where I sit against a peat-bank. They only notice me at almost touching distance, the fleeing otter exploding out of the way and back into the sea. The pursuer jumps back, stands about eight metres away, watching the other go, occasionally glancing towards me, uttering a soft wickering noise. Then he, too, runs down and into the sea again, and both totally disappear.

Between females from the same group range there was some animosity sometimes, but rarely more than a few calls, and resulting in mutual avoidance (Fig. 3.8). Only once did I see an encounter between females from different group ranges in Shetland, and that was a much more aggressive affair, with loud vocalizations and a chase. It could be, therefore, that females from strange territories are just as belligerent towards each other as males. Even between males and females there is often some aggression, especially from the female, but no actual fights. On mainland Scotland in 1991, a wild, large male otter broke into the enclosure of our old female at the Institute in Banchory, by climbing over the high wall. They had a vicious fight in which both got badly injured on legs and face, especially the old female. However, encounters between Shetland otters are relatively rare (except within families), given the numbers of animals along the coasts there. Somehow the otters manage to stay out of each other's way without actually meeting, quite likely a result of their sprainting habits.

When they do meet, there is very little visual display, and virtually no changes in facial expression, if any at all, just as Duplaix had recorded for other species of otters (Duplaix-Hall 1975; Duplaix 1980), and I had found for badgers. Probably this is a rather general phenomenon for mustelids, in contrast with the behaviour of, for instance, most canids (for example Tembrock 1957) and felids (Leyhausen 1956). In otters scent and sound play a more important role in communication; this is evident when two individuals meet on land, and usually the animal which is approaching will circle and come in from down-wind, checking the scent of the other one.

There was one type of behaviour which we saw fairly often, and which could perhaps be classified as a visual display: this was the posture in which large resident males 'patrol' the coasts of their territory. It was termed 'strutting', although the animals were, of course, swimming along: they were quite conspicuous, keeping parallel to the shore some 5 to 10 m out, exposing more of themselves above the surface than otters do normally, especially their tails. Usually when a male swam like that he would not be feeding, but would cover quite long distances, landing here and there for a spraint.

Our European otter is far less vocal than the South American giant otter, which was studied in detail in Guyana by Duplaix (1980). Here the sounds of screams, wails, barks, and explosive snorts accompany the troops of these animals everywhere on their forays along rivers in the rain forests: this is an elaborate system of messages, with many intermediates between the various calls, and allows for an almost infinitely detailed communication between the otters. There was no such articulation in the otters in Shetland, nor in any other species of otter I have studied so far. Thus it appears that, in otters, sound communication is much more elaborate in a species which lives in cohesive groups than in the more solitary species. A parallel evolution of systems of calls was observed in hyaenas, where the gregarious spotted species is by far the most vocal (Kruuk 1972, 1975; Mills 1990).

The few different calls which one does hear of the common otter usually carry over only short distances. There is one curious melodious exception, the *whistle*, which is the contact call between mother and offspring. When I first heard it, I could hardly believe that it came from a mammal, and I am sure that many naturalists must have heard that clear single note, and registered it as the call of a meadow pipit or some other bird. It carries over several hundred metres, and it is somewhat shorter and higher pitched if it is made by a small cub, compared with the call of its mother. It is also the call which one hears most often. Another one is the *huff*, which is hardly a 'call' but rather an explosive exhalation of air when an otter is alarmed by a person or a possible predator. Typically in Shetland, I would be working on the shore along a voe, emptying a fish-trap or something, and suddenly I would hear a soft huff behind me, only to see a ripple when I looked round. Females frequently huff in reaction to people when there are cubs.

The other calls are rather rare, and restricted to occasions when otters meet and one is aggressive. If an animal is cornered by another during a fight, it will produce a very cat-like *caterwauling*, but before things get as bad as that, there are calls better described as '*wickering*' or 'chittering', which are high-pitched, rattling sounds, often grading into a high wailing. Those are the sounds of a young intruder which has been surprised by a resident, or of a rough play-fight between cubs. But none of these calls are as loud as those that a noisy cat or a fox can produce.

The tendency of adult females living together in the same coastal territory in Shetland is to avoid rather than to confront each other, and this was demonstrated clearly on the few occasions when two families met, females with their cubs, because the cubs have no such scruples. The following is an account of two such observations:

October 1985. On a small rocky skerry in the Boatsroom Voe of the Lunna Ness, two small cubs are eating a big rockling, whilst the mother, Diamond, is fishing nearby. They are in the centre of their core area, within the group range. Suddenly another female from the same group, Nipple, seemingly from nowhere, arrives with her big cub which enthusiastically throws itself at the fish and the two smaller cubs. Some loud wickering and quick dashes back and forth, then somehow all three cubs eat, right next to each other. The mothers, however, are having more serious problems. Nipple sits downwind from the party of cubs, well out of the way and slightly crouched, not coming any nearer. Diamond in the meantime has caught another rockling and approaches with the red glittering prize, landing upwind of the cubs four metres away. Normally she would have taken the fish directly to the cubs; now she looks at the eating party, stands a moment, then slides back into the water fish and all, off to another rock out of sight. After a couple of minutes of status quo, Nipple's cub looks up, then walks to the water's edge; Nipple is there too and both swim off. Ten minutes later the two families are several hundred metres apart again.

Early November 1985. Following Nipple in the Boatsroom Voe, I see her with her single cub, quietly swimming between some small rocky islands in her own core area. She lands on one of them, obviously unaware that Diamond, from the same group range, and her two offspring are curled up between boulders on the other side. When Nipple comes downwind of the other party, she crouches slightly, hesitates, then slowly walks up. The two adult females sniff each other fleetingly, sniff anal regions, both in a low posture, circling warily but without overt aggression. In the meantime the cubs are running about in a mêlée of brown bodies and tails, two walk up to their respective mothers and try, unsuccessfully, to suckle. The females both spraint, close to each other. This chaos only lasts about 40 seconds, then Diamond slides into the sea with her cubs close behind her. Nipple and her cub continue for a short time sniffing the site of confrontation, then she spraints again and also leaves, fast, in the opposite direction.

There is no real aggression between these females here, they are probably related and their ranges overlap totally. But they keep out of each other's way, using their own preferred areas within the group range and away from the others. The system appears to be based on each otter having an interest in staying away from others, within their joint group range. Only the cubs do not seem to bother one way or the other.

SEXUAL BEHAVIOUR

In all the years in Shetland, we only saw five copulations of otters, all of them taking place between the end of February and the end of May. Mating always takes place in the water, and I think that it is only in captivity that otters may be more or less forced to copulate on land, giving rise to the notion that they may do it anywhere (Estes 1989; Pechlaner and Thaler 1983). The following describes one of my observations:

Early May 1986, along the Yell Sound, Shetland. In the late afternoon I meet Whisky, an otter bitch, in her core area. She is clearly recognizable from the bottle-shapped throat patch. I find her diving about ten metres out from the rocks, catching the odd eelpout and stickleback; she lands leisurely at a group of boulders. Suddenly she is all alert and tense, plunging back into the water—but now followed by one of our large males, with a yellow ear-tag. He catches up immediately, and for several minutes the two roll, splash and tumble about in an eruption of white bubbles. Yellow-tag grabs Whisky's neck and holds on under water, clasping her chest and curling around her. She briefly utters the whickering call, then rolls around so much that only small bits of otter show on the surface. The two quieten down with the male holding on, and occasionally they lift their heads for air. Seven minutes later they split up and both land between the boulders, briefly out of sight. Then I see Yellow-tag again, he spraints, lies down and grooms, next to the largest of the rocks. Ten minutes after that Whisky swims off, again catching the occasional eelpout or stickleback.

Several times we saw females copulate when still accompanied by cubs, quite grown up by then but still a very tight family bunch. The cubs mill around the mating pair or hang about on land nearby, obviously rather apprehensive, perhaps just somewhat scared of the male.

The mating sequences which I saw, such as the one above, may have seemed the product of a chance meeting. However, there was more to it than that—often, perhaps usually, the two animals knew each other well. In the winter before, I had twice seen Yellow-tag and Whisky meet, and there must have been several more occasions. These meetings were very brief, with the male approaching, sniffing, and the female usually somewhat offensive or even aggressive, then the male swimming off again. Bluebeard, a large old male who often came through the range of Baboushka, one of our females, sometimes played with her, and once I saw him catch a rockling whilst she was near, which he took up and left for her to eat. Such behaviour often occurs in birds as 'courtship feeding', but I was hesitant to read much into it in this case; I only saw it once and it could have been some chance occurrence. Later in that same year Andrew Moorhouse saw Bluebeard and Baboushka mate, in March.

There is no doubt that, in general, the males regularly meet up with the females in their ranges, and that the animals almost certainly know each other individually before it is spring, the mating season. I saw meetings between males and females on 34 occasions (Fig. 3.8); only in spring did this result in copulation. Sometimes meetings between male and female are highly aggressive, as is also suggested by the observations of Green et al. (1984), and in the previous section.

However, mating itself is a far less aggressive affair in this species than, for instance, in sea otters, where males bite the face of females during copulation, and may cause extensive injuries or death (Riedman and Estes 1990). Also in mink, females may be left with large wounds in the neck after copulation (Dunstone 1993); I have never seen such love bites in otters.

Generalizing from our small amount of information, the common otter appears to be probably polygynous as well as polyandrous: males overlap with ranges of several females and probably mate with them, and for each female range there are several regular male residents. In some form or other, this appears to be the common pattern for mustelids (Powell 1979; Sandell 1989; Kruuk 1989a).

It is also interesting to compare some of the observations on reproduction with what is known of the same species elsewhere, and of other otters. First, there is the seasonality. In Shetland the large majority of births take place in the middle of summer (see p. 228), and the copulations which we saw all took place in spring, with a known gestation period of 61–74 days (Wayre 1979). However, further south, in the rest of Britain and many areas of the continent, there is no seasonality of births, or if present it is not very strong. Mason and Macdonald (1986) state: 'the female otter appears to be continually polyoestrus, that is, there is a continuous oestrus cycle, with no specific breeding season'. In Shetland otters are probably seasonally polyoestrus, mating just at the time when life is toughest in that environment (Kruuk *et al.* 1987; see also p. 230). Questions arise as to what sets off the oestrus cycle in otters so differently in different parts of its range, and how their reproductive physiology is adapted to environmental requirements. We still have no answers to these problems.

Another question emerges from the comparison of reproductive systems between the American river otter and the European species. The river otter often has a delayed implantation (references in Melquist and Dronkert 1987), that is the embryo does not develop immediately after fertilization but may be 'dormant' for up to 10 months, then follows the usual 2 months development before birth. This occurs in a number of mustelid species and other mammals (review in Mead 1989). The effect of it is that the female river otter can mate immediately after giving birth to a litter, or later at any other time of year, and always produce cubs early in spring. Why should the two species, living at the same latitudes, in such apparently similar habitats and with similar ecology, have such different physiological mechanisms to cope with environmental requirements? There is no satisfactory functional explanation, as is true for many other cases where delayed implantation has been observed (Mead 1989).

THE FAMILY: PARENTAL BEHAVIOUR

With mating, the dog otter's reproductive effort ends: he has nothing more to do with provisioning or protecting the female and the cubs, it all falls on the female. This is what happens in almost all mustelids (Powell 1979); only in species such as the giant otter (Duplaix 1980) and the European badger (Kruuk 1989a) do males stay in regular contact with the cubs, although they are not actively involved in providing and protecting, and although there is no certainty of paternity. One often reads of a 'pair' of otters inhabiting a stretch of water, calling up pictures of happy family life: this is an anthropomorphism, and there is nothing in it as far as otters are concerned (although there is anecdotal evidence of a dog otter feeding young: Macaskill 1992). On the contrary, females may be aggressive to males when they come too close to young cubs (also in captivity) (Hillegaart et al. 1981), even if the male happens to be the cubs' father (as far as we know). The same observations have been made in other species such as L. canadensis (Melquist and Hornocker 1983) and L. perspicillata (Desai 1974). The aggression of the female appears to be part of a strategy to protect the young, at the same time as coping with their provisioning as efficiently as possible (see below). It is likely that male otters are a potential danger to cubs: males of many different species of carnivore have infanticidal tendencies (Packer and Pusey 1984).

Over the years we slowly pieced together the sequence of events when a female otter produces a litter in Shetland. The cubs are born in a 'natal holt', with a simple, very inconspicuous entrance (Moorhouse 1988). This natal holt is almost invariably quite far from the sea, sometimes almost a kilometre, sometimes only 100 or 200 m—but well away from other otter activity. It will have its own supply of fresh water, and the entrance will not show the usual pile of spraints (Green et al. 1984; Kruuk 1992). If one knows what to look for, the natal holt is recognizable because of the smooth earth around the entrance, which is larger than that of a rabbit hole, often with some dried mud on the grass outside and with the tell-tale footprints inside. Nevertheless, it still looks a very unlikely place for an otter, being so far from the shore. For about 2 months the cubs stay there, they are never seen outside; the mother is with them all the time except for one single foraging trip every day. She is highly secretive, at least nearby the holt, and when swimming along the coast nearest to the holt she hugs the shore closely, hardly showing herself. She feeds the cubs by lactation, and only at the end of the 2 months will she occasionally take a small fish to them.

In freshwater areas, natal holts may also be far from open water, as much as 1 or 2 km (Taylor and Kruuk 1990). However, L. Durbin (personal communication) found one natal holt in which one of his radio-tagged otters gave birth in a slope of boulders, close to a small tributary of a main river. Sometimes the animals may construct 'natal couches', which are elaborate structures of reeds or other material, collected by the females and fashioned into large, dome-shaped nests, with one or more entrances in the sides. These, too, tend to be far from open water (Taylor and Kruuk 1990).

One day, 8 to 10 weeks after their birth, the cubs follow the mother down to the water, the mother sometimes carrying them all or part of the way (Fig. 3.9). From then on the family will live mostly in holts close to the shore, frequently moving house, with the cubs regularly accompanying the female on her fishing trips (Fig. 3.10). It is likely that many of the natal holts reported in the literature (see, for example, the review in Harris (1968)) are, in fact, occupied by a family only after their first move, from the natal holt proper.

In the beginning of their life in the open, the cubs will mostly wait on the shore whilst the mother dives and then brings fish to them, and sometimes she 'teaches' them to fish. In our Shetland observations, the first immersion of a cub in the water is often forced, with the mother diving whilst carrying the cub by the scruff—but after a few days, the cubs seem as much at ease in the water as on land. They are still rather

FIG. 3.9 Female carrying cub from a natal holt down to the shore.

FIG. 3.10 Female followed by cubs during foraging.

clumsy, and being fat and fluffy they are much more buoyant than the adults. All this activity is made conspicuous by frequent whistles between mother and cubs; I have often thought that if there were predators around such as white-tailed eagles, wolves, or dogs, the young otters would be very vulnerable. In fact, one of our Shetland cubs was killed in this early period by a dog from the next door farm on Lunna.

Throughout the Shetland winter otter families stay together, the cubs often playing together (Fig. 3.11), slowly beginning to catch some fish themselves. In otters on the west coast of Scotland this happens when they are 4 to 5 months old (Watt 1993), but it may be somewhat earlier in Shetland. Usually when they are about 10 months old the family breaks up, some cubs leave earlier, but some stay much longer with the mother. At least, this was the case in Shetland (Kruuk *et al.* 1991), but on the Scottish west coast cubs stay longer with the mother, until they are 16 months or more (Watt 1993). Families of *L. canadensis* disperse when the cubs are between 8 and 12 months old (Melquist and Hornocker 1983). I had the impression in Shetland that it is mostly the mother who takes the initiative in the break-up; suddenly she is off, and if there are several siblings they stay together somewhat longer in the same area where they were reared. Some cubs show streaks of independence at an early age, often fishing apart from the others, and leaving for good well before the rest split up.

FIG. 3.11 Cubs playing in the sea.

There are many variations on this family theme, and this is no more than a general picture. When the cubs are small and fed only by suckling, they are kept away from other otters, being protected almost full time by the mother in a well-hidden location. Although this can be far from the feeding areas, it is without too much cost to the mother, as she only makes one foraging trip a day. After a couple of months the strategy changes, because (a) the cubs are less vulnerable to intraspecific aggression and predation, and (b) weaning starts, so the mother is then carrying fish to the young, one at a time. At that stage provisioning alters, as the cubs stay close to the female along the shore line, and they are fed near to where the prey has been caught.

Several stages in this sequence of events are important in the population ecology of the animals. The first is inside the natal holt, when the cubs are dependent solely on their mother's milk for their survival. Unfortunately we know next to nothing about what happens down there, about suckling behaviour or competition between siblings, and about conditions when food is scarce. These are issues I would like to know much more about, after we had found the close relationship between food availability and the numbers of cubs during that time of the year (Kruuk *et al.* 1991; see also p. 230).

During the first days of the cubs' life outside the natal holt and when weaning starts, we have seen several times that otter mothers deliberately abandon some cubs, leaving them to starve (p. 233). It is unclear why, if females are going to abandon any of the cubs, they do so at this stage rather than much earlier inside the natal holt. One can only speculate; perhaps the abandonments which we actually watched were only the tail end of a process which mostly takes place underground. Or perhaps it is easier for the mother to abandon a cub during the course of ferrying them individually over long distances, rather than inside the holt where they are together in one bunch. Or perhaps the female avoids having a decomposing corpse in the natal holt. The abandonments are not 'accidental', nor are the abandoned youngsters in obviously inferior condition.

By the time that the families are swimming along the Shetland shores, usually during late August and throughout September, it appears that, whatever factor it is which decides the numbers of cubs, this has already taken its toll. From the families which we observed, only a few cubs disappeared in the following months, and it seems that once a cub is about 3 months old it is relatively safe. There is a steady build-up in the numbers of families in the autumn, and I found that in an average winter, on almost half of the occasions when I saw otters in the Lunna study area, they were animals in families (Figure 8.13, p. 230).

Once the families are moving about along the shores, with the cubs accompanying the mother, they can be seen active at any time during daylight; at night they are in one or other of the larger holts near the shore. During the day the female is very busy keeping herself and the cubs fed, as these field observations describe:

Early February 1987, Lunna Voe. Weibka, a young female, fishes 80 metres out, whilst two six to eight months old cubs, her first litter, clamber between the big boulders on the steep shore. I am within a few metres, peering down on them over a large lump of stone. Whilst I am absorbed by the playing cubs, there is a splash pretty close by. Weibka is on her way, with the water churning around her: she has an enormous black fish which struggles wildly. Slowly, with great effort, she gets the prey to her cubs.

She lands a conger eel, an animal longer than she is herself, and maybe also heavier. She drags it onto the rocks, well away from the water; it has almost stopped struggling. As soon as Weibka releases the fish, one of the cubs starts eating, tearing at it somewhere in the middle. The other cub sits close by and sniffs Weibka; both groom and roll around together. Five minutes later, Weibka is off fishing again, whilst one of the cubs eats and the other sits nearby. After ten minutes, the cub which is not eating enters the water and follows Weibka, swimming around where she is diving. Weibka comes up with a small fish, an eelpout, and quickly eats it herself on the surface, then dives again. This time when she emerges she has a bullrout: almost 20 cm long, with a conspicuous red, white-spotted belly. This is a plump, large fish and a substantial meal; Weibka heads for the shore again, deep in the water, and with a large bow-wave caused by the fish in her mouth. The cub follows and Weibka lands close to her conger eel. The first cub is still eating there, and the second one gets the bull-rout. Weibka just sits, grooms herself and rolls over, some ten metres away, mainly out of sight from the cubs.

November 1985, Boatsroom Voe, Shetland. A female with one four months old cub is fishing just off-shore, she in the water and the cub on land. The mother catches one eelpout after another, all of similar length, between 15 and 20 cm, eating all of them herself in the water, floating back-up between dives, chomping loudly on her prey and ignoring the cub. Then, 20 metres out, she captures a big squat bullrout, probably some 18 cm long, much heavier than any of the eelpouts. Immediately she makes a straight line for the shore, dropping the fish next to her cub. The youngster just looks at it lethargically, clearly totally satiated already. The mother sits and waits, watching the cub, but when nothing happens she eats the bullrout herself, and both curl up on the seaweed.

In the first of these observations the conger eel was an unusually large prey—in fact, I stole the remains after the otters had finished with it and ate it myself. Both cases were a demonstration of one of the principles of otter family life: when the female is providing for her cubs, she eats small prey herself, and takes the larger fish to her offspring (Fig. 3.12). It makes good sense in energetic terms—the food for the cubs has to be ferried over much longer distances than the food for the female and she can only take one item at a time.

In my observations in Scottish freshwater lochs there was no difference in the size of prey eaten by the cubs and that eaten by the mother because virtually all prey was relatively large (eels), and they were almost always taken ashore. Only rarely did I see otters eat fish on the surface of the water there. But when otters feed on a mixture of small and large fish in rivers and streams, the same maternal behaviour as in Shetland may well apply; good observations are difficult to come by because of the nocturnal habits of the riverine animals.

When the cubs are beginning to catch some fish for themselves at the age of about 5 months they are initially much less efficient at this than their mother, and they continue to improve their fishing skills well into their second year. This has been studied in detail by Jon Watt, on Mull, off the Scottish west coast (Watt 1991, 1993). He found that even though cubs would be totally self reliant and independent from their mother (on Mull this would be at the age of about 13 months), they still were less adept at catching fish and continued to improve their percentage of successful dives, and the mean weight of prey they did manage to catch. Not until they were 15 to 18 months old were they as efficient as an adult in terms of diving success and proportions of time spent under water and on the surface. The prey which young cubs caught themselves was often slow and easy, and of rather low quality (low calorific value);

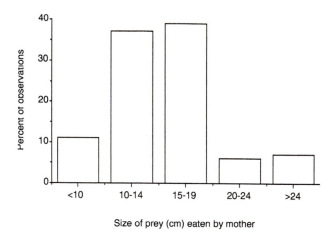

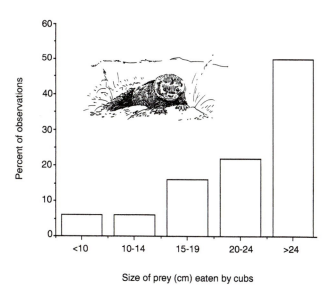

FIG. 3.12 Prey sizes eaten by hunting mother otters in Shetland (number of observations $n = 103$), or taken by them to their cubs ($n = 22$): $\chi^2 = 49.0$, d.f. $= 2$, $p < 0.001$. (After Kruuk *et al.* 1987.)

for instance, they captured many shore crabs, which have the lowest food value for effort of all types of prey recorded. Watt found a negative correlation between the age of cubs and the proportion of crabs in their food; in his study area crab accounted for 32 per cent of the prey captured by dependent cubs, but for independent sub-adults this was 15 per cent, and for adults only 3 per cent.

It is tempting to assume that otters have to learn their fishing skills, although the improvement of performance with age may have nothing to do with actual learning, it could just be maturation. It certainly has the appearance of learning, with the cubs watching closely when their mother is fishing, often diving with her from the age of 4 or 5 months onwards in beautiful synchrony. Sometimes they may be peering down from the surface of the water, floating along, watching the mother at work on the bottom. This behaviour is similar to that recorded from fox cubs, when the vixen is catching earthworms (Macdonald 1980*b*).

But the most suggestive observations of 'learning' are those where the otter mother appears to actually teach her cubs to catch prey such as the following:

September 1986, midday, at low tide along Shetland's Yell Sound. In the kelp nearby an otter called Mrs. Fitz is diving and rummaging about; below me, no more than twenty metres away, is her single cub, a chubby little animal, about three-quarters of its mother's length and probably about three months old. I had just watched a long series of dives by the mother which had lasted several hours.

Then Mrs. Fitz heads for shore, carrying an eelpout. She lands close to the cub, shaking herself whilst still holding the fish in her mouth. With the little one begging alongside her, she walks up to a small rock pool, no more than 20 centimetres deep, and about 2 by 1 metres in size. She drops the eelpout into this; it is an average size fish, less than 20 centimetres long and still well alive, wriggling off into the pool. The cub goes after it clumsily, and misses. Mrs. Fitz comes to the rescue and deftly picks up the eelpout again, releasing it within seconds just in front of the cub. At the second attempt the cub is more success-ful—perhaps the eelpout has slowed down a bit. Mrs. Fitz stands motionless for about half a minute, watching her cub eat.

Before the cub stops feeding the mother returns to sea, back into the huge heaving mass of kelp. She again emerges with an eelpout, a smaller one this time, of less than ten centimetres. She quickly chews this up herself before getting another, one of about the same length as the one she had taken to the cub. She lands it, takes it to the pool, and the whole sequence repeats itself. The fish is released, the cub chases it and misses, the mother catches the fish and releases it again, then the cub's second chase is successful. After the cub has bolted down the eelpout, mother and son play, rolling about on the rocks and in and out of the pool; finally both dive into the sea at great speed and swim off next to each other, closely along the shore, into the huge landscape of the Yell Sound.

I have seen this behaviour in several different females, and it must be a fairly common occurrence in the life of otters. It reminded me of earlier days in the Serengeti, in East Africa, where I had seen cheetah mothers return to their clumsy little cubs and drop a live baby gazelle in front of them, leaving the young ones to chase it but catching it for them when they could not cope. Would such 'teaching' really contribute to the hunting skills of the next generation? Watt (1993) saw otter cubs themselves release prey close to the shore, then catch it again, sometimes several times with the same fish, and it has been described for mongooses as 'prey capture play' by Rasa (1973) (see also reviews in Fagan (1981) and Bekoff (1989)).

I saw otters 'playing' with fish a few times in Shetland and once on the Scottish west coast. This usually concerned solitary, subadult, independent otters landing and releasing a fish close to the shore, recatching it several times, and mouthing and pawing at it whilst lying or floating on their backs. They roll with it in the water, sometimes throwing the fish right up into the air—this could go on for up to 10 minutes. Interestingly, in Shetland, the prey with which otters played was usually a flatfish (the top-knot, when I could distinguish the species). That was not a very common prey, a less preferred species, and one that was sometimes discarded again when an otter caught it after the animal appeared to be more or less satiated. Lumpsuckers are also used in play.

In fresh water, otters sometimes appear to 'play' with eels they have landed, rolling around and flinging them about, even taking them into the water again. This may be functional in removing slime from the prey, and the eels are always eaten afterwards.

Watt (1993) suggested that the period of a year or more that otter cubs are dependent on their mothers is unusually long for a carnivore of this size, and comparison with equally large sympatric species such as the fox (becoming independent when it is about 4 months) (Macdonald 1980*b*), badger (4 months) (Neal 1986; Kruuk 1989*a*), and wild cat (3 months) (Corbett 1979) bears this out. Watt relates this to the long learning period that otters need to acquire hunting skills. Catching fish may be unusually difficult, a theme which returns repeatedly in these studies of otter ecology.

4

Diet

The prey of coastal otters in Shetland was determined from direct observations rather than faecal analysis, and it is compared with data from different coasts and other studies, from otters in fresh water, and from other species. It consisted almost only of fish, mostly small, eel-shaped, bottom-living species. But otters also took some larger prey, especially in winter, and particularly the males as well as females with cubs (who preferentially fed larger fish to their offspring). Occasionally birds and rabbits were taken. The larger fish were important in terms of proportion by weight in the diet, again mostly in winter. Prey were smallest and most diverse in spring, and least varied in summer. There was no substantial variation in the diet of individual otters fishing along the same shores. Feeding and diving efficiency improved with age during the first 18 months of life. In fresh water otters ate similar size fish as in the sea, with similar differences in diet between the sexes. Otters consume a quantity of fish of at least 15 per cent of their body weight per day. The diet of other otter species is described; there are broad similarities world-wide, but where different species occur sympatrically there are striking divergences.

INTRODUCTION

The food of otters is the one aspect of their ecology which has been studied quite thoroughly, in many different places (see, for instance, the summary in Mason and Macdonald (1986)). This is not so much because that is what everyone is interested in, but because it is a more accessible problem than, say, social behaviour or population ecology. Diet can be studied by analysing the otters' faeces, the 'spraints', and these are easy to find, often on conspicuous sites in convenient places, such as under bridges.

Otters are highly specialized animals compared with other carnivores. One can express this, for instance, by comparing the data on prey categories of otters from different areas, with similar data sets from other Carnivora. For this we can recognize several prey categories such as fish, birds, small, medium-sized, and large mammals, amphibians and reptiles, vegetable matter, and several others. We then calculate K, the Kendall coefficient of concordance, which expresses the agreement in the occurrences of these prey in diets of the same species from different places; K will range from 0 to 1 (Siegel 1956). If a species is relatively opportunistic, its diet will tend to vary between sites, there is little agreement between sites, and K will be low. Taking published data on diets from a sample of carnivores, then otters were the most specialized, with $K = 0.82$ (data from 12 different studies), compared with $K = 0.44$ (fox, 11 studies), 0.67 (wild cat, 6 studies), 0.77 (lion, 7 studies), and 0.81 (badger, 12 studies) (Kruuk 1986).

Although such comparisons between diets from different areas may be useful, for any study such as ours in Shetland or on the Scottish mainland, the question should be asked whether it would really be worthwhile to add even more data to what we know already about otters' food. This especially if it meant spending months in the laboratory with stacks of spraints whilst some of the information was already available. The answer had to be affirmative; firstly, there were only a few studies which considered the food of otters against the background of prey populations available. Secondly, we were interested in whether the otters' food supply was important to their numbers and dispersion, so it had to be looked at together, and simultaneously, with other aspects of otter life.

This chapter will concentrate on our own studies of food in Shetland and in the rivers of north-east Scotland. In Shetland we were in the fortunate position of being able to watch otter predation at work, otters with prey in their mouths, and this enabled analysis of many more aspects of foraging than one could do with the customary faecal analysis (Kruuk and Moorhouse 1990). The information on diet could then be related to availability and productivity of prey populations (place and time of capture), to differences between individuals, and otter social and foraging behaviour and we could investigate the role of food as a possible limiting factor in populations, and the way in which it could affect the predators.

However, for our observations elsewhere, on the Scottish mainland, where otters were active at night in streams and lochs, we had to revert

to the usual faecal analysis, relating results to data on the prey populations (Carss *et al*. 1990, Kruuk *et al*. 1993).

The main questions about otter food which first needed to be addressed in the Shetland study were:

1.   Which species, and what sizes of fish do the otters take?

2.   Are there differences in the diet at various times of the year, and
     along different parts of the coast?

3.   Do all otters take the same prey?

The answers can then be related to the availability of various fish and
crabs (Chapter 5). This will make it possible to assess how foraging behaviour, feeding strategies, and their timing, are adapted to food availability (Chapter 6), and finally all these factors can be brought together
to show the effects of food supply on otter populations (Chapters 8 and
9) and their social organization (Chapter 9).

Some earlier information on the species which otters eat in
Shetland was available from Herfst (1984), who analysed spraints of
animals in our study area at Lunna, and from Watson (1978), studying
spraints from otters on the island of Fetlar. Other scat analyses from
sea-living otters were done on the Scottish west coast (Mason and
Macdonald 1980, 1986). Most of these earlier data were collected in
the summer; they showed consistently that otters take relatively small
fish, mostly bottom-living species, as well as varying numbers of crabs.
But it is difficult to generalize much further from these results
because what was available to the animals was not known, and the
data were not complete for all seasons. Watt (1991), using faecal
analyses and direct observations, related the food of otters at all times
of year to prey availability along the coast of Mull, in western
Scotland. He calculated differences in the profitability of different
prey in terms of time spent by the otters, and calories gained from
various types of food.

Outstanding research on otter diet and foraging in fresh water includes the early work by Sam Erlinge, in southern Sweden (Erlinge
1967), the PhD study by Margaret Wise in the south-west of England
(Wise *et al*. 1981) and the project in western France by Libois and
Rosoux (1989). All these projects compared the remains of prey in the
spraints of otters with indices of prey availability, with numbers of fish
caught in nets, and with electro-fishing. The main conclusions were
similar: prey other than fish is not important, and fish are taken more or
less according to availability, with a preference for the slower-moving
species, and often for the smaller fish in the populations.

Recently in rivers in north-east Scotland we were able to estimate diet from faecal analysis, and total food intake from both field observations and data on captive otters (Carss *et al.* 1990; Kruuk *et al.* 1993). These data were used together with information on otter numbers and fish productivity to look at the overall relationship between otters and fish populations.

METHODS OF DIET ANALYSIS

The common method for obtaining information on the food of otters is through the analysis of their spraints, identifying the prey remains. Many parts of the skeleton of prey are left undigested by the otters, and in particular the vertebrae of fish can be identified down to species (methodology and keys by Webb (1975), Watson (1978), and Conroy *et al.* (1993)), and they can be used to estimate the size of prey eaten. There are some major problems in the interpretation of the information from spraint analysis, because the extent of digestion of fish bones and other remains depends to some extent on species and size of the prey, and one needs correction factors to overcome this (D. Carss in preparation). Also, one usually does not know which otter left a particular spraint, where the prey was taken, and when.

One of the advantages which we had with our otters in Shetland was that we could see what they were doing, so that we could identify what species they came up with, the size of the prey, exactly when and where it was caught, and by which otter. When an otter was foraging, it swam close to the coast with its head, and sometimes part of the tail, showing above the water. In one quick smooth movement it would go down almost vertically, lifting its tail out of the water before disappearing (Fig. 6.1). After perhaps 20 or 30 seconds the otter reappeared close to where it went down. The chance was one in three that it would come up with a fish in its mouth, usually of a small manageable size which it could eat then and there floating on the surface, pointing its snout upwards and occasionally guiding the fish with a forepaw (Figs 6.2, 6.3 (p. 157)). Larger fish, or prey which was difficult to handle, would be taken ashore (Figs 4.1, 4.2, 4.3, 4.4).

When the animal was eating we could identify it from the throat patch; we could often see what species of fish it had caught, and estimate how large it was. If a dive was unsuccessful, the otter would go down again after a few seconds of recovery on the surface. Otters do not swallow prey under water; both in captivity and in the wild I saw them actually catching prey under water on numerous occasions, and it was always eaten on the surface.

FIG. 4.1 Young male otter eating a small crab.

FIG. 4.2 An otter eating a sea-scorpion (*Taurulus bubalis*) next to exposed kelp (*Laminaria digitalis*).

Identifying the prey at a distance needs some experience. When we were studying the otters feeding, we were also operating a concurrent scheme of trapping fish (Chapter 5), so we handled fish almost daily. There were only a limited number of species which were relevant to the otters, so it was quite feasible to learn to distinguish them even at a dis-

FIG. 4.3 Yearling male otter eating a shore crab (*Carcinus maenas*).

FIG. 4.4 Male otter eating a dogfish (*Scyliorhinus canicula*), along exposed Shetland coast with thong weed (*Himanthalia elongata*).

tance of 100 m or more, with binoculars or a telescope. However, it often happened, of course, that we could not be sure what species of fish an otter was eating, or we could only see that it was 'something eel-shaped' or 'pollack-shaped'. This was a problem, especially with very small prey less than 10 cm long; nevertheless, about half of what the otters came up with could be clearly identified. We estimated the size class of prey by comparison with the width of an otter's head, which is about 8 to 10 cm near the eyes.

It was possible that the high proportion of unidentified prey in our observations caused a bias in the results. However, we feel that this bias was unlikely to be large, because even if prey species could not be identified, we could often estimate size, and whether, for instance, the fish was eel-shaped or not. These data supported the assumption that unidentified fish were broadly similar to the identified prey.

To convince ourselves that we could recognize these fish in the jaws of an otter, we did some small experiments. One of us, the observer, would sit with a pair of binoculars at about 70 metres from another person, the experimenter, who would be manipulating different fish in the way that an otter would. Seventy metres was a usual distance for otter observations. From a large collection of dead fish, caught in traps in the otters' range, one would be picked up, fingered, and partly hidden in one hand, turned over and manipulated, just as an otter would do when eating, for a period of 5 s, which was less than an otter would take. Both the observer and the experimenter wrote down the species of fish, then the experimenter would record the measured size, and the observer the estimated size. All of us were fairly consistently almost 80 per cent correct in our assessment of species, and with the sizes we were about 12 per cent out, in a total of 112 presentations.

The experimental situation of our observation-check was only an approximation of otter-reality. We only used binoculars, and in the wild we also used a telescope; in the experiments we worked with dead fish from the deep-freeze, which had often lost much of their original bright colour, and we could not use the characteristic movements which different fish make when caught by an otter. Also, when the observer wrongly identified a species in the experiments it was often in confusion with a very similar species—we did not, for instance, mistake a saithe for an eelpout, but we might report it as a pollack. The results were encouraging enough for us, therefore, to continue with the assessment of diet by direct observations, rather than from faecal analysis.

When we saw an animal foraging, we would record the coastal section where it was, distance from the shore, and the time of day. Apart from identity of the individual otter, and behavioural information (other otters present, what happened before and after), we would record prey species and size class: class 0 was a prey smaller than 5 cm, class 1 was between 5 and 10 cm, class 2 between 10 and 15 cm, and so on. From the

relation between length and weight (Nolet and Kruuk 1989), we estimated the weight of the prey.

The statistical analysis of the results presented some problems.
Although we collected large numbers of observations, a total of about 45
individual otters were involved, and most observations were collected
from fewer than 20 animals. I had to be careful, therefore, about 'dependent observations': if just a few of the otters had some peculiar idio-
syncrasies, such as a preference for a particular kind of prey, then this
could have had a considerable effect on the conclusions. This problem is
often overlooked in diet analysis, but usually one cannot do anything
about it. In Shetland we could recognize individual otters when they
were feeding; wherever possible, therefore, I used individual otters as
the unit of observation in the analysis. For instance, when comparing
the importance of eelpout in the diet in spring and in summer, I compared the proportions of eelpout in the diet per individual otter in
spring with the individual proportions of eelpout in summer. This does
not affect the overall figures presented, but it does mean that if there is
a statistically significant difference in some comparison it is not caused
by just one or two otters being awkward but by more individual otters
behaving in a way which is characteristic to them.

RESULTS

Diet in Shetland

In Table 4.1 all species are listed which we recorded as part of the otter
diet. This includes only those observations where an otter was actually
seen foraging, and not those cases when I came on an otter already
eating on the shore. This often happened with large prey such as lumpsuckers (Fig. 4.5), bright red fish which took a long time to consume, or
dogfish (small sharks; Figs 4.4, 4.6). It was less often that we saw otters
catch any of these big fish, and I had to be careful not to distort the diet
record towards large fish—as it was, they were an insignificant, though
spectacular, part of the list of fish species caught. Apart from the lumpsuckers and dogfish we also found otters with large cod, ling, conger
eels, octopus, and various others, though more rarely. Lumpsuckers
were an interesting prey, because of the sexual discrimination by the
otters: they caught almost only males. On only one occasion in many
dozens of observations did I see an otter with a female (which is bright
green, in contrast to the male which is red). Presumably, lumpsucker

TABLE 4.1 Fish species taken by otters foraging in Shetland, all observations (but including only those observations in which otters were seen before the capture of prey)

Species	No of observations	Per cent
Eelpout *Zoarces viviparus* (L.)	686	33.8
Rocklings *Ciliata mustela* (L.) and others	343	16.9
Sea scorpion *Taurulus bubalis* (Euphrasen)	283	14.0
Butterfish *Pholis gunnelus* (L.)	201	9.9
Stickleback *Gasterosteus aculeatus* L.	114	5.6
Saithe *Pollachius virens* (L.)	106	5.2
Bullrout *Myoxocephalus scorpius* (L.)	73	3.6
Pollack *Pollachius pollachius* (L.)	45	2.2
Saithe or pollack	9	0.4
Flatfish (mostly *Zeugopterus punctatus* Bl.)	39	1.9
Lumpsucker *Cyclopterus lumpus* (L.)	18	0.9
Ling *Molva molva* (L.)	16	0.8
Cod *Gadus morhua* (L.)	6	0.3
Eel *Anguilla anguilla* (L.)	5	0.2
Poor cod *Trisopterus minutus* (L.)	5	0.2
Pipefish *Syngnathus acus* (L.)	4	0.2
Sea stickleback *Spinachia spinachia* (L.)	4	0.2
Dogfish *Scyliorhinus caniculus* (L.)	2	0.1
Wrasse *Labrus mixtus* (L.)	2	0.1
Conger eel *Conger conger* (L.)	1	0
Yarrel blenny *Chirolophis ascannii* (Walb.)	1	0
Other fish	3	0.1
Shore crab *Carcinus maenas* L.	61	3.0
Starfish *Asterias rubens* L.	1	0
Rabbit *Oryctolagus cuniculus* (L.)	3	0.1
Total	2031	99.6%
Unidentified prey	1557	43.3%

males are more vulnerable because they guard the eggs, an interesting disadvantage of parental behaviour (Clutton-Brock 1991).

Figure 4.7 shows the contribution of each prey species to the otter diet throughout the year, in two different ways. In terms of numbers caught, the eelpout (Fig. 4.8) came first, and eelpout, rockling (Fig. 4.9), sea-scorpion (Fig. 5.2, p. 130) and butterfish (Fig. 5.1, p. 129) together

FIG. 4.5 Male lumpsucker (*Cyclopterus lumpus*), 1.2 kg, taken by an otter. Otters rarely prey on the female of this species.

FIG. 4.6 Characteristic otter damage to dogfish (*Scyliorhinus caniculus*) (only the liver is eaten).

made up about three-quarters of the prey. But in terms of weight the picture was different; rocklings were most important, and the four species mentioned earlier accounted for less than half of all food. I will say more about the common, small fish species in Chapter 5.

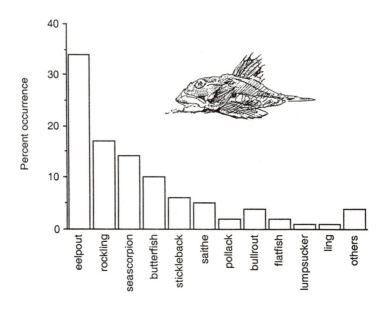

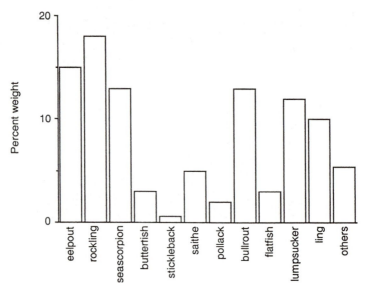

FIG. 4.7 Different species of fish in the diet of otters along the Shetland coast, in terms of numbers and of estimated relative weight. Based on 2031 identifications (1557 fish were unidentified).

FIG. 4.8 Important prey for Shetland otters: the eelpout (*Zoarces viviparus*).

FIG. 4.9 The five-bearded rockling (*Ciliata mustela*).

Clearly, the large, less frequently taken species of prey made quite an important contribution to the diet, with big fishes such as bullrout (Fig. 4.10), lumpsucker (Fig. 4.5), and ling accounting for about 35 per cent of the food in weight, but for only about 5 per cent in terms of numbers.

FIG. 4.10 Bullrout (*Myoxocephalus scorpius*), a larger relative of the sea-scorpion.

We also saw otters catch birds (fulmars, *Fulmaris glacialis*, and a guille-mot), and several more rabbits and other really large fish, apart from the ones mentioned in Table 4.1. However, those were rare events, incidental occurrences which were not recorded during regular foraging observations, and we probably noticed them only because they concerned such big and spectacular prey. They are therefore not comparable with the other food data. Predation on sea birds was also seen in sea otters in California (Riedman and Estes 1988), but again as occasional events and insignificant in terms of otter diet. In general, therefore, prey consisted of fishes that were either slow swimmers, or eel-shaped, or both. Most of them were small: the median size of prey was 16 cm or 28 g. The distribution of prey sizes is plotted in Fig. 4.11; it shows the unexpectedly high contribution of 'very large' prey to food intake in terms of weight.

Otters ate remarkably few crabs in Shetland (almost always when they did, it was the common shore crab)(Figs 4.3, 4.12, 4.13). They were only a minor constituent of the diet, therefore, although crabs were exceedingly common along the coast, only a few individual otters would eat them (see below). The results of spraint analysis by Herfst (1984), of material collected in summer over a much wider area around the Yell Sound, generally supported our conclusions. When I returned to our study area several years later, in 1993, I saw otters catch crabs much more often than before, and along the coast there were many spraints with crab remains. This was quite a striking difference from before, and it coincided with unusually low fish populations (Figure 8.15, p. 232).

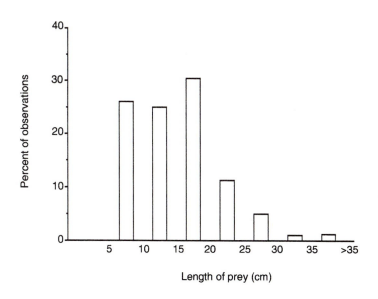

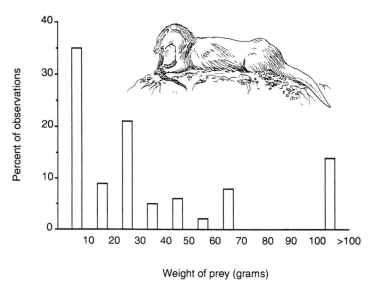

FIG. 4.11 Top: distribution of estimated lengths of prey caught by otters in Shetland. Bottom: distribution of weights of prey, calculated from lengths (using data from fish caught in traps). $N = 3523$ observations.

FIG. 4.12 Remains of shore crabs (*Carcinus maenas*) left by an otter.

FIG. 4.13 Characteristic otter damage to carapaces of shore crabs (*Carcinus maenas*),
showing indentations of canines at the rear end.

Seasonal changes in food in Shetland

It was necessary to know what otters were eating in different seasons
because there could be seasonal fluctuations in the availability of populations of certain prey species. For simplicity, Fig. 4.14 shows just the
annual changes in the proportion of numbers of each prey caught, not

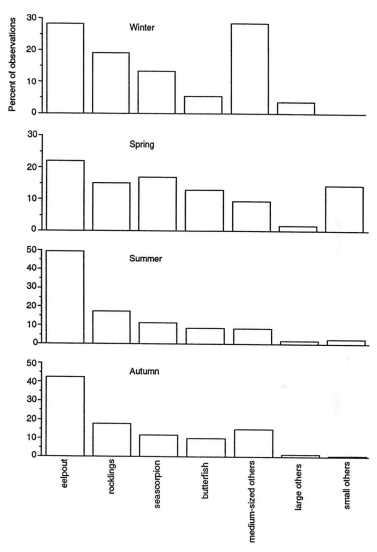

FIG. 4.14 Proportion of prey species in the Shetland otter diet during different seasons. Number of observations: spring (March–May) = 754, summer (June–August) = 434, autumn (September–November) = 479, winter (December–February) = 361. Statistical significance of seasonal differences: χ^2 = 521.7, d.f. = 3, p < 0.001.

changes in the weights. Although at a glance the overall picture looks rather the same throughout the year, there were some quite remarkable changes in the diet, which were of obvious importance to the otters. Eelpout were, at all times of the year, more numerous in the food than any of the other species, but they were more than twice as frequent in the summer diet as in the spring: in summer, and to a lesser extent in autumn, they dominated the food. Rocklings and sea-scorpions fluctuated little, at all times they were a steady 15 to 19 per cent and 12 to 17 per cent of the diet respectively. The other species varied more markedly; it was especially striking that in spring the small light-weight and/or low calorie prey (butterfish, sticklebacks, crabs) were prominent. In winter the heavier saithe and pollack were more common, as well as several heavy 'other species' such as ling, lumpsucker, bullrout, conger eel, and cod (Kruuk and Moorhouse 1990).

In general, otters in spring had a more varied diet, evenly spread over many species with much low-quality food. In summer they specialized by eating large numbers of fairly light prey, which was mostly of one species only and of better quality than in spring. In winter the diet was not as varied as in spring (but more so than in summer), with more large, heavy fish being taken than at any other time. The autumn diet was neatly intermediate between that of summer and winter, with a gradual transition away from small-fish specialization to the winter variety of larger fish.

The differences between the seasons suggested that there was quite substantial variation in mean and median sizes of prey throughout the year. Means and medians turned out to be rather different, mostly because the few large fish which otters took had a big effect on the mean weight of prey, but very little on the median. Mean weights of prey expanded from spring, through summer and autumn to winter, from 46, 49, to 61 and 61 g, a significant increase ($\chi^2 = 10.9$, d.f. $= 3$, $p < 0.02$); median weights were 22 g in spring and 28 g during the rest of the year. These were the first indications from the otters' food that spring was a hard time for otters.

The last point will become more obvious when we take into account the otters' fishing behaviour, and the behaviour and movements of the fish. But before that, it is important to consider the diet of otters in different places, along the various stretches of coast of the study area. This complicated matters as regards seasonality: otters fished in different places at different times of year. These variables in the otters' diet interacted with each other, but first I will discuss each of them separately.

Diet along different shores

The food which the animals were eating in different places could well affect the social organization, the extent and use of home ranges by otters all year round, as suggested in the Resource Dispersion

Hypothesis (Macdonald 1983). We collected relevant data on both diet and food dispersion.

The coast of Lunna Ness was rather a mixture of different types of shores. At one extreme there were the beautifully sheltered bays, the 'voes', often separated from the wild world outside by a sand bar, with only a small opening to let the tides through. In contrast there were also steep cliffs, highly exposed, with huge, wild waves breaking against them—and in between these extremes there was a gradation of coasts of different exposure. The east side of the peninsula which was our study area (Fig. 2.3, p. 33) was high, with the bottom of the sea falling off steeply, to about 80 m deep close inshore. The west side was a shallow coast, intersected by the voes, and one had to go a long way out to get as deep as 20 m. The otters used these coasts differently: this showed in their foraging behaviour, as I shall show later, but it was evident also in the kinds of fish which they obtained there.

Clambering along the steep, rocky slopes of the east coast, it was much more difficult to see the animals, for various reasons. The otters behaved differently, the water was choppier, and looking down on it from high up the cliff there was less chance of seeing a small head cutting the surface than from down along the shore. But if we did see an otter fishing along the east coast, the chances were much greater than elsewhere that the animal would come up with a really large prey. Somehow it fitted in with the landscape when an otter parted the waves with a large red lumpsucker in its mouth, or with a cod or a flatfish the size of a pan-lid. A mere eelpout would have been out of place.

Figure 4.15 shows the otter food along different coasts of Lunna Ness which were ranked in order of 'exposure'. This ranged from small, highly protected bays to voes with relatively large openings to sea, then carried on along the somewhat sheltered west side of the peninsula further north up the coast which was progressively more affected by the large open waters of the Yell Sound, until finally reaching the totally open, wild east coast.

There were interesting trends in the relative frequencies of species which the otters took along those shores. Perhaps the most important effects were the changes in the importance of eelpout and the heavier five-bearded rockling with increasing exposure. This was obviously relevant to the otters' use of a particular type of coast, as eelpout and rocklings were the two most significant prey species. To emphasize this phenomenon, in Fig. 4.16 the percentages of eelpout are plotted against rocklings in the diet for different coasts: in areas where otters caught many eelpout (the sheltered voes), they fared badly with rocklings, and vice versa. Various other, large prey species, such as bullrout,

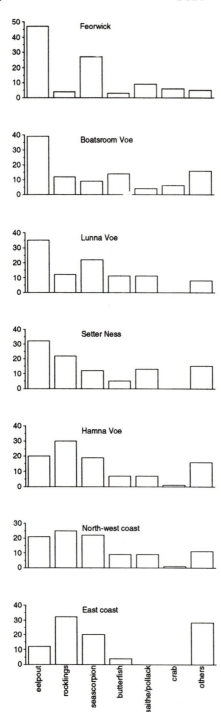

Fig. 4.15 Prey species taken by otters along different coasts in Shetland, with wave exposure increasing from top to bottom, from Feorwick (sheltered small bay) to Boatsroom Voe and Lunna Voe (larger bays) to the open coasts, with the East Coast extremely rough and exposed. Numbers of observations from top to bottom: 79, 886, 233, 465, 201, 127, 25. (Data from Kruuk and Moorhouse 1990.)

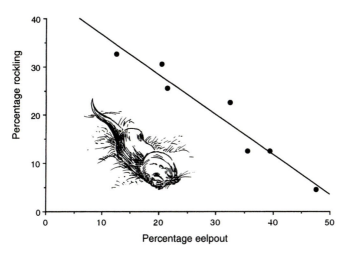

FIG. 4.16 Correlation between the occurrence of eelpout and rockling, in the diet of otters along different coasts. Data from Fig. 4.15. Correlation coefficient $r = -0.96$, $p < 0.01$ (regression $y = 43.8 - 0.83x$).

lumpsucker, ling, and flatfish, were more likely to be caught by otters along the exposed shores. In general, therefore, the more exposed a coast, the larger and heavier the average prey size for otters: a phenomenon which was caused largely by otters taking different prey species, but Fig. 4.17 shows this just in terms of estimated weight, irrespective of species.

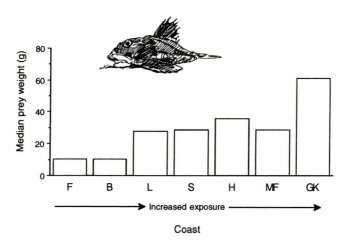

FIG. 4.17 Size of prey of otters along different coasts with increasing exposure (see legend to Fig. 4.15). Statistical significance: $\chi^2 = 191.5$, d.f. $= 6, p < 0.001$.

For statistical analysis, I first treated all observations as if they were totally independent of each other; the trends were highly significant (Fig. 4.17). Next the coastal differences in the food of each of 13 individual otters were analysed, and this showed exactly the same pattern as in Fig. 4.17, which was again highly significant. Almost all animals ate larger prey the more exposed the coast along which they were feeding ($\chi^2 = 9.3$, d.f. $= 1, p < 0.01$), and they showed the same pattern for proportions of eelpout and rocklings taken.

Can one decide whether the differences in diet between seasons were caused by the otters changing their feeding sites? With data such as these, a neat multivariate analysis is not appropriate (because of interdependence of observations within individual otters) and I had to make do with surveying the data for broad patterns, in which both seasonality and movements along coasts play a role.

For instance, of all the observations on otters catching prey in our study area, in spring and summer the majority were in the large, sheltered Boatsroom Voe (B). In autumn we saw them fish about equally often in the Boatsroom Voe and along the more exposed coast of Setter Ness (S), and in winter more than two-thirds of our prey-catching observations were along Setter Ness (Fig. 4.18). It was difficult to allow for differences in our observational effort; but basically we were with the otters, wherever we found them, so Fig. 4.18 gives a reasonably accurate representation of the seasonal shift in otter foraging, from sheltered areas in summer to more exposed coasts in winter.

This annual move to other feeding grounds could be related to the availability of fish (see Chapter 5), but in the proportions of different kinds of prey which were caught in these places in various times of year there were only few clear overall trends. They were obscured partly because the sample sizes were getting rather small when I broke them down into seasons by section. One trend which was outstanding was that the rag-bag of 'large' prey, ling, cod, conger, and others, increased substantially in the diet towards the winter and only along the exposed coasts. Similarly, the collection of 'small' prey, sticklebacks, crabs, butterfish, and so on, increased in the diet in spring and this happened much more in the sheltered voes than elsewhere. Eelpout increased substantially in summer, and this increase took place especially in the sheltered voes. The percentage of rocklings in the prey remained almost the same everywhere, at all times.

Generally, the main trends were that otters moved into the voes and sheltered coasts in spring and summer, where they fed first on 'small' prey, and later in summer mostly on eelpout. In autumn there was a move to more exposed coasts where in winter they did most of their for-

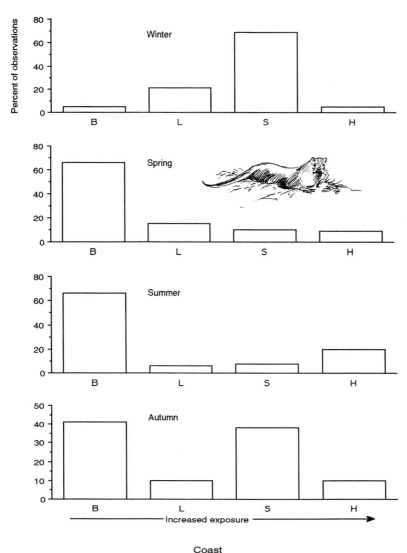

FIG. 4.18 Frequency of our feeding observations of otters along different types of coast in the centre of the study area, over the seasons: **B** = sheltered bay (Boatsroom Voe), **L** = less sheltered bay (Lunna Voe), **S** = rather exposed open coast (Setter Ness), **H** = most exposed open coast (Hamna Voe). One observation = one otter feeding bout: $n = 349$ (spring), 330 (summer), 291 (autumn), 200 (winter).

aging, taking many larger prey species (which they did not catch along those coasts at other times). The animals appeared to follow the shifts in prey populations over the seasons, along the various shores within their home ranges.

Diet of individuals

The pattern of diets which change with the seasons and along different shores may be complicated, but the complexity goes further still. Males tend to live more along exposed coasts than do females, and this sexual difference also fluctuates with the season (see p. 41). Before discussing this further, some overall differences in the food of the various categories of otters should be mentioned.

Figure 4.19 summarizes the main species of prey taken by males, females on their own, and females which were provisioning cubs. The differences were small and not significant when analysed with individual otters as the sampling unit. But the various categories of otter did select substantially different prey sizes. First, in the observations over the whole period of the study the largest fish were taken by males, next came females with cubs, then females without cubs, then independent cubs, and the smallest fish were the diet of the unknown otters (Fig. 4.20). The differences in median prey sizes were, however, not as large as the difference between means. For instance, the mean weight of fish caught by males was 62 g whereas for all females it was 53 g, but the median weights were 28 g in both cases. The mean weight of prey of males was 13 per cent heavier than that of females, and this difference between means and medians suggests that it was due largely to differences between the heaviest fish caught, and not to males selecting heavier prey right across the spectrum.

Females without cubs caught fewer large prey than mother otters did. The prey which males took was 22 per cent heavier than that of females without cubs, but there was very little difference between fish caught by mothers and fish caught by males. The mothers themselves ate the same sizes of fish as did the other females, but they also caught larger prey, which they took to their cubs (Kruuk *et al.* 1987).

For statistical evaluation prey weight was ranked for otters of different status, and within seasons and within coastal sections ('voe' versus 'exposed') to avoid bias. Males and females with cubs scored the same ranks of prey size, but the other categories had significantly lower scores ($\chi^2 = 10.5$, d.f. $= 4$, $p < 0.05$).

The next question was whether individual otters had their own feeding preferences, their own idiosyncracies. For example, when two females fished along the same coast during the same season, did they take the same prey? I did have some data on this, but of course one is not likely to get a massive number of observations in such detail. In the Boatsroom Voe we watched both Mrs Fitz (F11) and Diamond (F44) in the same area during one summer and autumn period, and along the

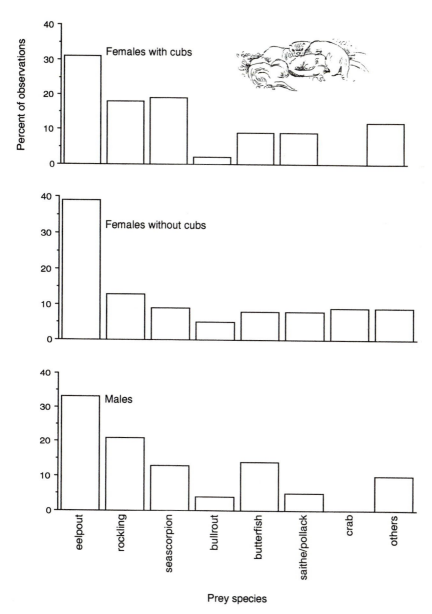

FIG. 4.19 Food of males, single females, and females with cubs (numbers of prey observed 510, 660, and 638 respectively). The differences are statistically not significant when using individual otters as the sampling unit.

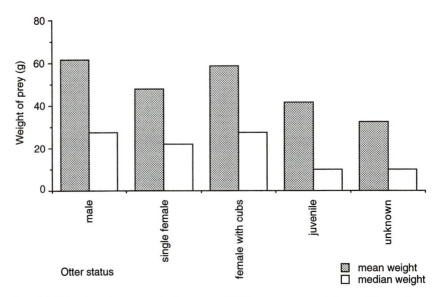

FIG. 4.20 The size of prey caught by otters of different status: median and mean sizes. See text. Numbers of observations: male 809, female 917, female with cubs 946, juvenile 108, unknown status 315. Statistical significance: $\chi^2 = 10.5$, d.f. $= 4, p < 0.05$.

Setterness Coast, Baboushka (F7) and Weibka (F16) were often seen fishing in exactly the same area during one autumn and winter. Within these two pairs of females, there was a remarkable similarity in what they brought to the surface, with differences of only a few percentage points (Fig. 4.21).

However, during the same summer when we were watching the two females in Boatsroom Voe, there were also two newly independent subadult otters feeding there, a male (M37) and a female (F38). They showed a striking contrast, although they often operated in exactly the same sites (Fig. 4.21). Three quarters of the prey of the young male were crabs, the rest were small, three-spined sticklebacks which lived in the brackish water close to one of the burns entering the voe. The young female did not take any crabs, only fish, and of those half were stickle-backs, the rest mostly butterfish. None of these prey species were popular with the adult otters. It was not likely that the difference between these two young otters had anything to do with their different sex, because adult males and females in general had similar prey species preferences. There were genuine individual trends here, just in the newly independent animals, and from the other comparisons one might expect that these differences would disappear again when they grew older.

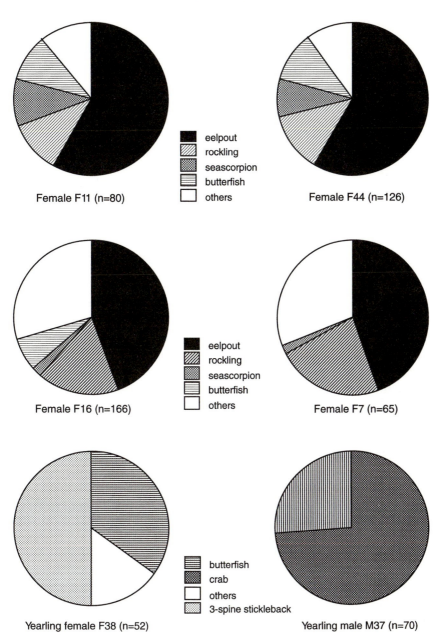

FIG. 4.21 Diets of three sets of individual otters which were observed feeding during the same season along the same coasts: females **F11** in same area as **F44**, **F7** in same area as **FI6**, and yearlings **F38** in same area as **M37**. Differences within the sets of adults were negligible, but significant for the two young otters (overall $\chi^2 = 78.1$, d.f. $= 3, p < 0.001$; see text).

The phenomenon of young, independent otters taking relatively low-value prey for several months after independence has been documented in detail by Watt (1993). He showed that in otters on Mull the proportion of crabs in the diet decreased from about 50 per cent in cubs of 4 months old to about 10 per cent at 1 year of age, and it was almost zero for adults. There were no significant differences in the proportion of various species amongst the fish which these animals took when they grew older, perhaps because on Mull there was less choice of fish species, and most of the prey was smaller and of lower calorific value than in Shetland (p. 148).

OVERVIEW: THE FOOD OF OTTERS IN SHETLAND AND ELSEWHERE

In other studies almost all information on otter food has been gathered from spraint analysis, as in our own work in north-east Scotland. The Shetland observations confirmed results elsewhere, which show that almost uniformly the vast majority of fish taken by the otters are relatively small with a median length of 13 cm. Watt (1991) established that the most common prey fish in the otter diet in the sea around Mull were butterfish of about 16 g, and sea-scorpions of about 30 g. Wise *et al.* (1981) found that salmonids taken by otters in streams in Devonshire, England, were around 12 cm long (median) and eels 25 cm, that is 21 g and 22 g respectively. In Poland the median length of fish taken by otters (mostly Cyprinids) was less than 10 cm, weighing less than 20 g (Brzezinski *et al.* 1993), eels taken by otters in the west of France had a median length of 26 cm and a median weight of 25 g (Libois and Rosoux 1989), and in the Dee valley in north-east Scotland they took eels of 27 cm median length, weighing 29 g, and salmonids of about 10 cm median length, which were around 15 g (Jenkins and Harper 1980). Our own results from the River Dee and tributaries showed salmonids in the otter diet with a median length of about 8 cm (Kruuk *et al.* 1993; see Fig. 5.11), except in winter (see below).

In Ireland freshwater crayfish (*Austropotamobius pallipes*) made up about half of the diet of otters, and of the fish species available there mostly perch (*Perca fluviatilis*) were taken, with a median length of 10 cm (weighing 10 g) (Le Cren 1951). The Irish otters also ate many frogs, salmonids with a median length of 16 cm (49 g), and eels of 35 cm (weight 72 g) (Kyne *et al.* 1989). For further detailed lists of otter food species and references see Mason and Macdonald (1986).

Although most prey is small, at times large fish are also preyed upon, and at least in Shetland the occasional lumpsucker, bullrout, cod, or ling makes a much greater contribution to the otters' food than appears at first sight, especially in winter. This is a phenomenon which would be easily overlooked if we assessed diet from spraints alone, because with large fish disproportionately few bones are consumed. Along rivers in Scotland, such as the River Dee, otters may take many fish of well over 30 cm in winter (Fig. 4.22). For instance, in 57 otter-killed salmon found in some of the tributaries of the Dee the median length was 71 cm, with a median weight of 2.9 kg (Carss *et al.* 1990). At those times, large salmon are the main prey (Fig. 4.23), and otters eat on average 975 g from each fish in one session (285–2075 g). There was some evidence that it was mainly the large male otters which took big salmon, again a parallel with the Shetland observations.

Another general characteristic of otter prey is that most species are slow-moving and bottom-living, at least at the time of capture (Mason and Macdonald 1980; Herfst 1984; Kruuk and Moorhouse 1990); this will be discussed in more detail in the next chapter. We never saw otters with any of the fast pelagic fish which occurred along our Shetland coasts, such as mackerel (*Scomber scombrus*), or even saithe and pollack during the seasons when these species swam in open water rather than between the weeds.

Some generalizations can also be made about the seasonal variation in diet. For instance, it was striking in the Shetland data that during spring otters ate much 'low-quality' food, such as crabs, sticklebacks, butterfish, and in general light-weight prey (Kruuk and Moorhouse

FIG. 4.22 Female salmon killed and partially eaten in characteristic manner by an otter, and left on a riffle in the River Dee, north-east Scotland.

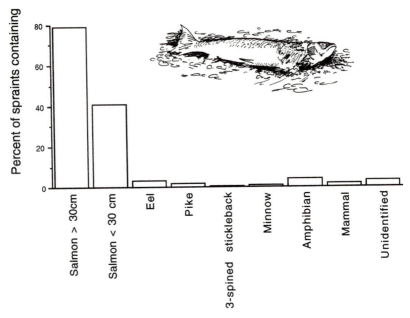

F<small>IG</small>. 4.23 Prey remains in spraints of otters along the River Dee and tributaries, north-east Scotland, from November 1989 to January 1990. Number of spraints = 324. 'Salmon' includes *Salmo salar* and *S. trutta*; 'minnow' is *Phoxinus phoxinus*. (Data from Carss *et al* 1990.)

1990). I will show later that this occurred in conjunction with other signs of food stress in spring.

In summer the Shetland otters had a less varied diet than at other times: a much greater proportion of their food consisted of just one species (eelpout). For badgers it has been shown that during times of plenty they had a more specialized diet, and that the food in sparse populations was more varied than that of badgers in the really high densities (Kruuk and Parish 1981; Kruuk 1989*a*). If a similar mechanism operated in the Shetland otters, this would suggest that the summer is a time of plenty for them, which was confirmed by fish-trapping (Chapter 5).

Our Shetland data showed differences in the size of prey in the diet of males and females, and in the size of prey with which females provision cubs and that which they eat themselves. In Scottish rivers we suggested similar differences between males and females, with larger fish (especially salmon) taken more often by males (Carss *et al.* 1990). Such variation in diet between sexes is not surprising, given the large difference in size (p. 42); moreover, it is common in many species of carnivore, especially in mustelids. It can go as far as in the wolverine (*Gulo gulo*), where males feed mostly on reindeer, females on rodents and birds (Myhre and Myrberget 1975). Honey badger (*Mellivora capensis*) males regularly kill small ungulates and jackal-sized carnivores, but the females subsist on small rodents, scorpions, and reptiles (Kruuk and Mills 1983). The subject has been reviewed by Moors (1980) and Ralls and Harvey (1985), who concluded that the biological function of this sex difference in food was not the avoidance of competition, but a by-product of the difference in size between male and female. This size variation would be favoured by evolution, because small females need less energy for daily maintenance, so that they can channel more energy into reproduction.

The fact that female otters carry the larger prey to the cubs makes good sense energetically. Otters can only transport one fish at a time, and sometimes they have to do this over considerable distances from the shore. There is little point, then, in carrying a stickleback.

QUANTITIES CONSUMED

Although much is now known about the kinds of prey which otters consume, actual quantities taken and the variations therein are far more difficult to estimate, and one has to rely to a large extent on observations made on captive otters. Two otters kept by Stephens (1957) each ate on average about 1 kg of fish per day, over a period of 8 months, which was approximately 15 per cent of their body weight per day. Wayre (1979) weighed the food for two captives over 1 week, and found daily consumptions of 12.2 and 12.8 per cent of body weight. Our own captive otters, which were kept in an open enclosure with a large swimming pool, produced rather similar estimates. For instance, one adult male (9.5 kg) maintained his body mass over 3 months in summer on 1.13 kg fish per day ($\pm$ 0.24 kg s.d.; $n = 89$ days), which is 11.9 per cent of his body weight per day. Over the same period a neighbouring group of two females and one male (6.5, 8, and 10.5 kg) also maintained body weight by together eating 3.15 kg per day ($\pm$ 0.29 kg), or 12.6 per cent of their combined body masses.

In the wild consumption is more difficult to assess. We estimated food intake by one large male along a tributary of the River Dee in the northeast of Scotland, which we could follow and observe intensively in winter by radio-tracking. He ate almost only large salmon and we could weigh the left-overs and measure the length of the fish, from that we could estimate its original weight and thus calculate food consumption of the otter. This 8.0 kg animal ate on average 975 g from every fish caught ($n = 39$), and as he took one salmon every night this amounted to 12.2 per cent of his body weight per day (Carss *et al.* 1990). This was an underestimate of daily consumption, as the otter also took some other small prey which we could not assess as accurately.

In Shetland we followed a radio-tagged lactating female (with one cub) and estimated her total intake per day, from total intake per feeding period and length of time fishing per day (Nolet and Kruuk, in press). Over a 12 day period, this 5.4 kg animal ate 1.5 kg of fish per day (± 0.2 kg), or 28 per cent of her body weight. This would have been a higher consumption than that of an average otter, as the animal was lactating, nevertheless these figures suggest that the estimates based on observations in captivity may be too low. This is not unexpected, as the energy consumption in captivity is likely to be far less than in the wild. For otters in nature, a daily food intake of 15 per cent of body mass per day is a conservative estimate, and this will be used in later calculations, ignoring the likely seasonal variation in consumption.

DIET OF OTHER SPECIES

We have at least some basic information on the diet of almost all species of otter, though generally this is less detailed than what we know of our *Lutra lutra*. It is interesting to look at this information against the background of their geographical distribution; there are otters on all continents, except in Australia and Antarctica.

Each continent has at least one (exceptionally two or even three) species of otter in its freshwater areas which is predominantly a fish-eater, specializing in rather slow, often bottom-living species (Foster-Turley *et al*. 1990). In Europe it is *L. lutra*, in south-east Asia the smooth otter (*L. perspicillata*), as well as *L. lutra* (Wayre 1974, 1978; Kruuk *et al*. 1993). Little is known of *L. sumatrana*, the hairy-nosed otter. In Africa the spotted-necked otter (*L. maculicollis*) feeds almost exclusively on fish (Rowe-Rowe 1977; Kruuk and Goudswaard 1990), and similarly in North America the river otter (*L. canadensis*) (Melquist and Dronkert 1987). In South America the neotropical river otter (*L. longicaudis*) is piscivorous, as is the much larger giant otter (*Pteronura brasiliensis*) (Duplaix 1980).

Next to these fish-eaters, there are also freshwater species which specialize in crustaceans, mostly crabs (and some also specialize in molluscs although they may eat fish). In south-east Asia this is the small-clawed otter (*Aonyx cinerea*) (Wayre 1974, 1978; Kruuk *et al*. 1993), and in Africa from the same genus, in different parts of the continent, the clawless otters (*A. capensis* and *A. congica*) (Rowe-Rowe 1977; Kruuk and Goudswaard 1990). In South America the southern river otter (*L. provocax*) consumes mostly crustaceans (Chehébar 1985). Europe and

North America do not have any otters which are freshwater invertebrate specialists, probably because of the absence of crabs. The otters there readily take crayfish which have only a localised distribution.

Finally, there are the otters which live exclusively in the sea, also specializing on invertebrates, crabs, and molluscs: the sea otter (*Enhydra lutris*) in Pacific North America (Kenyon 1969; Riedman and Estes 1990), and the marine otter (*L. felina*) in Pacific South America (Ostfeld *et al.* 1989). Elsewhere, various freshwater species of otter may also use the sea whenever conditions permit, as does the European otter in Shetland.

Whenever several species of otter occur sympatrically, there is a clear difference in diet, although there may be some overlap. For instance, off the coasts of Alaska the sea otter fishes in deep water, taking many molluscs, echinoderms, and crustaceans as well as fish (Kenyon 1969), and the river otter catches prey close inshore, mostly small fish (Melquist and Dronkert 1987). In Africa we found the spotted-necked otter and Cape clawless otter along the same coasts of Lake Victoria, with the first eating exclusively fish and the second almost exclusively crabs (Kruuk and Goudswaard 1990).

Most spectacularly, in Thailand we found three species sharing the same rivers and in regular contact with each other, the small-clawed otter eating almost only crabs, the common otter taking small fish and amphibians in almost equal proportions, and the smooth otter concentrating on large fish (Kruuk *et al.* 1993). There was some evidence there that the common otter was partially excluded from the lower reaches of the river by the larger smooth otter, and it lived more in the tributaries and higher up the main stem of the river, where there were more frogs.

Clearly my superficial description of the diet of otters world-wide is incomplete, but it suggests a simple dietary pattern underlying the geographical dispersion of species, with competitive exclusion. Apart from otters, other species of carnivore are also a part of this scene; in Africa the water mongoose *Atrilax paludosa* occurs in many places where there are otters, eating mostly crabs (Skinner and Smithers 1990), and similarly in south-east Asia the crab-eating mongoose *Herpestes urva* shares its habitat with the three otter species there. In this last case we could demonstrate that the mongoose eats significantly smaller sizes of crabs than the small-clawed otter (unpublished observations), but with a large overlap between them.

5

Fish as prey: numbers, behaviour, and availability

Most of the otter prey species in Shetland were least active by day, and at low tide, when they were hiding and were therefore easiest to catch for a foraging otter. Butterfish and shore crabs were more common than eelpout and rocklings, which were the species taken most by otters. In Shetland otters selected the larger individuals from prey populations, but elsewhere they took non-selectively from the bottom-living prey species. The main prey species in Shetland lived in different, well-separated habitats along the coast, and this was likely to affect the size of ranges which otters needed to exploit such prey. There was a glut of potential prey in mid-summer, and fish were scarce and relatively small during late winter and spring. Sites from which we repeatedly removed all fish were repopulated within a day, providing otters with places for sustained harvesting. Predators may have major effects on communities of prey species, and otters took a high proportion from populations of salmonid fish in Scotland, but the effect of otters on fish populations along the Shetland shore was probably small.

INTRODUCTION

To fully understand the relationship between otters and their food one would have to watch otter–fish interactions in every detail. The best possibilities for this exist in the clear waters off Shetland, with otters active by day, rather than in freshwater lakes and rivers of mainland Britain, where otters are nocturnal and the waters often murky. Several of the sea-fish on which otters feed have a very striking appearance, with subtle patterning which is sometimes brightly coloured; to see them in their environment proper, when scuba-diving in the breath-takingly beautiful clear underwater world of rocks and seaweeds, is an unforgettable experience. However, try as I might whilst diving along the Shetland coasts in a dry suit, I was never able to see wild otters find and take even one single fish. Quite simply, I just could not keep up with the otters in the dense forests of huge algae. I did see fish, many of them—but even in that I was nowhere near as efficient as the otters. In places where it took an otter 10 or 20 seconds to emerge with a fat rockling, I might be rummaging around for half an hour before I finally found one. The otters' performance was totally superior to mine.

If direct observations of the interaction between predator and prey were impossible, both in the sea and in freshwater areas, the next best thing were various indirect methods of deducing what happened between otters and fishes. There were several important questions:

1. Which species and which sizes of fish are there, and how does the food taken by otters relate to that (that is, what do they select)?

2. How do fish behave, for example where exactly in the habitat are they at various times of day, and what are they doing when the otters are hunting for them? What is their distribution, what happens after predators remove fish?

3. What is the fish density, the biomass, and the productivity in different zones and places, and throughout the seasons?

In the end we were able to get information in the Shetland study area on most of these problems (Kruuk et al. 1988), except for the one concerning productivity, which we could only measure in the rivers and streams of study areas in north-east Scotland (Kruuk et al. 1993). It is useful to discuss these fish observations in some detail, despite the fact that otter populations elsewhere will be dealing with totally different prey situations, and what we were seeing in our Shetland and Scottish study areas would only be relevant for the otters there. At least our data give an idea of the kind of relationship between this predator and its food species, and it is easy to see that the answers to the above questions would be of immense importance to the otters.

One problem initially was the embarras de richesse: there were so many species of fish, all fascinating in themselves. Even after more than 3 years of regular fish-trapping along the coasts of our study area in Lunna, Shetland, we kept coming up with new species in our traps: we caught over 30 species of fish, as well as various crabs and lobsters. Shetland's distance from the centres of scientific activity in Britain meant that little was known about the littoral fauna, so we had to start at the beginning and make an inventory, and get to know the species.

The fish which appeared to dominate the Shetland scene, in terms of the numbers we caught along the coast and as a prey for our otters, was the eelpout, *Zoarces viviparus* (Fig. 4.8, p. 105): a fascinating animal with various peculiar characteristics. It is highly camouflaged with its subtle pattern of green, brown, and orange; it has a body like an eel and a face like a frog, with bulging eyes and thick lips. Its bones, the bits one finds in an otter spraint, are sea-green. A typical specimen is less than 20 cm long, weighing 10 or 20 g; the largest we ever caught was 28 cm whilst the Shetland 'largest ever' (caught by an angler in Lerwick harbour) was 34 cm, though it can reach 50 cm (Muus and Dahlstrom 1974; Wheeler 1978). As the scientific name indicates, the eelpout produces live young, but very little otherwise is known of its behaviour. It is normally found under stones, close to the shore, and this, together with its eel-like shape and the young being found inside, suggested to German fishermen that the eel must originate in the eelpout: they called it 'eel-mother' (*Aalmutter*). In an aquarium or when diving at night, one notices that they do swim around in mid-water like other fish, but almost only in darkness: they are clearly nocturnal (Westin and Aneer 1987). Eelpout are very seasonal in their appearance along the Shetland shores; it appears that they usually live in somewhat deeper waters, but they come close inshore to mate and reproduce, especially in the summer. Eelpout are less common further south, and are almost absent around Mull (Watt 1991).

The other main actor on the fish scene for the otters was the five-bearded rockling, *Ciliata mustela* (Fig. 4.9, p. 105). There are some similiarities with the eelpout: it is somewhat eel-shaped, it is found under rocks and it is of comparable size (20–30 g, but up to 200 g in weight). But the differences are striking: *Ciliata* is dark-coloured, brownish-red on top, it has a much more slender head, and the sense organs are totally different. As the name suggests, this species has barbels which dominate the frontal appearance. There is an interesting organ, peculiar to rocklings, which they carry on their back just in front of the dorsal fin. This is a groove in which tiny fin-rays, a modified first dorsal fin, are in almost continuous motion creating a small current. This may be an organ for chemical detection, as suggested by Kotrschal *et al.* (1984). The five-bearded rockling is common everywhere along the European Atlantic coasts; its young are pelagic and their huge summer swarms are a frequent food for the smaller sea-birds. Even more than the eelpout it is a species of the night, and whilst one may see the odd eelpout around in daytime when diving or watching the aquarium, rocklings are strictly nocturnal.

The situation was complicated in Shetland by the occurrence of four species of rockling, of which the five-bearded was by far the most abundant. We did catch the other three species every so often, especially the large three-bearded rockling *Gaidropsarus vulgaris*, which could be over 500 g in weight, a rather ugly, blotchy, dark reddish-brown fish, a species of somewhat deeper water (Wheeler 1978), but in practice also an inhabitant of the Shetland shallows. The northern rockling *Ciliata septentrionalis* is small like the five-bearded species, similarly a fish of deeper waters (Wheeler 1978), but we caught it fairly often. It was distinguishable from the five-bearded by its somewhat larger head and curiously lobed lips. The shore rockling *Gaidropsarus mediterraneus* was also a rather big species, weighing up to 500 g; it was the rarest of the four.

The butterfish *Pholis gunnellus*, and the sea-scorpion *Taurulus bubalis* were very common (Figs 5.1, 5.2). These are the species one sees most when diving or snorkeling, apart from the very small (and for our purposes irrelevant) gobies. The butterfish are beautiful to watch, with their sharply defined black spots along the back, and snake-like move-

FIG. 5.1 Butterfish (*Pholis gunnellus*) at night.

FIG. 5.2 Sea scorpion (*Taurulus bubalus*).

ments. They occur in large numbers intertidally, in shallow waters
between the rocks, stones, and algae, but they also go down to deeper
waters, well out of reach of the otters (Wheeler 1978). However, they
are light (usually 5–15 g), and in shallow waters otters did not eat them
that often. We found that individual butterfish were far less likely to be
caught in our fish-traps than eelpout and rocklings. There is not much
flesh on a butterfish, even on a big specimen: they are curiously
flattened from side to side, they feel very slimy, and also spiny because
of the hard sharp rays in the dorsal fin. They are not much of a prize for
a predator therefore. Like rocklings, butterfish are mainly nocturnal
(Westin and Aneer 1987), but nowhere near as exclusively so.

In contrast, the sea-scorpion is active mostly during the day, if active
is the right word: most of its time is spent lying still on or between the
rocks and weeds, waiting for its prey to move. Short and broad, 5–50 g,
dark brown and with its head and gill-covers covered with bony knobs
and bumps and very sharp spines, the sea-scorpion should be well
camouflaged and protected against onslaught from an otter.
Nevertheless one sees them quite easily when scuba-diving, and otters
do not seem to be deterred by the spiny protection. They handle them
rather carefully, though, often taking even small specimens ashore to
eat. There is also a close and highly prized relative of the sea-scorpion

which otters take fairly often: the father-lasher or bullrout, *Myxocephalus scorpius*. It looks similar to the sea-scorpion, but usually it is quite a bit bigger, often weighing 60–120 g, and what is especially striking is its bright red or orange belly, with white polka-dots. It is a thrilling sight to see an otter carrying one ashore with the bright flash of red pointing ahead and the large pectoral fins standing out, sometimes obscuring the otter's forward view. This species is known to be nocturnal in summer and diurnal in winter (Westin and Aneer 1987).

These were the common fish which seemed to play a most important role in the life of an otter, but I could have included several more in this overview, such as the abundant saithe *Pollachius virens*, and the rather similar pollack *P. pollachius*, and even the ordinary stickleback *Gasterosteus aculeatus*. Also larger species such as lumpsuckers *Cyclopterus lumpus* and dogfish *Scyliorhinus canicula* occurred regularly and were popular with otters.

SHETLAND FISH: AQUARIUM OBSERVATIONS

In Shetland I was fortunate enough to find an experienced aquarist, who made us a glass fish-tank measuring $1 \times 0.4 \times 0.3$ m, which was perfect for observing fish in captivity. We installed it in front of a window in a quiet room in our bothy close to the shore, and for many months we kept in it several of the fish which interested us, mostly rocklings, eelpout, sea-scorpion, and butterfish. Obviously one has to be very careful in using observations of captive creatures to understand what is going on in the wild. But by using the appropriate combination of several species of fish, the local seaweeds (various wracks), and rocks, we were able to catch at least a few glimpses of life as it is being lived along the shores. For instance, it was immediately clear that rocklings, eelpout, and butterfish were very largely active only during the night (Fig. 5.3), especially the first two. These were data which I collected by just recording whether the fish were active or not at all times of day, without disturbing them, though if necessary I used a small torch.

When the fish were not active they were almost impossible to find in the aquarium, hidden as they were somewhere beneath a stone, or in the densest parts of the weeds, beautifully camouflaged and keeping absolutely still. As an experiment I tried several times to disturb their peace by stirring the weeds with my hand, in the way I thought an otter would have done with its head. The fish did not budge, and I could even touch them without difficulty; there was little doubt that an otter could have caught them easily if it had been able to distinguish the fish from the weeds by touch. On the other hand, when the fish were active (at

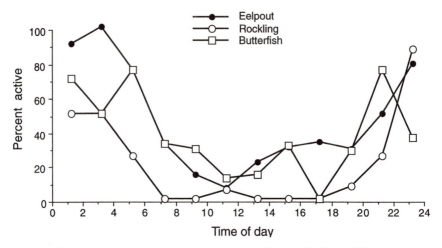

FIG. 5.3 The activity of 3 eelpout, 3 five-bearded rockling, and 4 butterfish in an aquarium, using natural light. Shetland, June 1986. Graphs show the proportion of fish present which were active during spot-checks over a period of 15 days.

night) they moved fast, and I was quite unable to get anywhere close with my clumsy fingers. If I had had to choose a time to catch these fish by hand, it would have been daytime—an observation which was clearly relevant to the otters' behaviour.

I did not get many observations in the tank on the sea scorpion, nor on the fifteen-spined or sea-stickleback *Spinachia spinachia*, but in general they appeared to be typical diurnal species, lying quietly on the bottom at night. The sea-scorpion, which is a frequent prey of the otter, never swam about a great deal in any case and could be touched by hand at all times. It was always easily visible, and it seemed to be the proverbial sitting duck for an otter.

After these incidental observations, and by snorkeling and scuba-diving, we had some idea about where and how the relevant fish occurred along the Shetland coast, and about the way in which otters could catch them. This needed to be quantified, to study the seasonal and other differences in fish availability, to describe the way in which the different fish used the coast and to get an estimate of fish densities. For this we had to rely on a more objective system of sampling the wild populations. We therefore used small, purpose-built fish traps, and the results from these traps turned out to be all-important for our understanding of otter ecology in Shetland.

FISH-TRAPPING IN SHETLAND

Methods

The principle of a fish-trap is simple: it is a container with a funnel-shaped entrance (Fig. 5.4). Ours were made of a proprietary brand of heavy-duty, plastic gauze called 'Netlon' which had holes of 7 mm diameter, separated by black strands of 3 mm wide. I had found this material, purely by accident, being used for road-making in Shetland where the peat is covered by large sheets of Netlon before gravel and stones are dumped on it.

Each trap was cylindrical, 50 cm long with a diameter of 27 cm, with one side of the cylinder consisting of a funnel ending in an opening of 7 cm. At the other end of the cylinder we made a small flap for removing the fish which we had caught; inside we tied a large stone which kept the trap on the bottom and prevented it from moving about too much. The trap was attached to a 20 m line, and on days when we were sampling fish we threw 10 or 20 of the traps from the shore into the sea, usually at low tide, attaching the lines to boulders or pegs. We never baited the traps, and we usually checked them to collect our catches a day after setting them.

Later we improved on the model by making the cylinder longer (80 cm), and adding a second funnel in line with the first one so that the trap had effectively two chambers, an entrance porch and a holding chamber. This twin-funnel type was more effective holding the rocklings, saithe, and pollack once they were caught. It meant we had to calculate correction factors so that we could compare the catches between

FIG. 5.4 Standard one-funnel trap used for sampling fish populations in Shetland.

the two models by setting both types of trap simultaneously, fairly close together (which we did 140 times). For instance, we found from such experiments that the twin tunnels caught five times more saithe and three times more rocklings than the single funnels, but the two types were equally effective for eelpout, and for shore crabs.

Trapping results: totals over four years

Combining the catches from 1493 'trap nights', from all over the study area and at all times of year, we caught 4141 fish and 1122 crabs, a total of 5263 (Table 5.1). Such results, however, could not be used as they stood for the purpose of assessing fish availability, for several reasons. The fish fauna was different along different types of coast, so the results depended on where we put the traps. Secondly, there were seasonal fluctuations for the various species, so the results were partly determined by the time of year when we did most trapping. Thirdly, and probably most important, there was variation in 'trappability' between species. I will try and unravel these complications in the following sections.

With these reservations in mind, Table 5.1 gives an overall idea of the species composition of the fish fauna, but it does distort. For instance, since butterfish are almost five times more difficult to trap than eelpout (p. 143), the results in Table 5.1 really show that butterfish were the next most common species present (there would be 36 butterfish for every 44 eelpout had they been equally trappable, not 8, as in Table 5.1). Similarly, a rockling is more than 10 times more likely to be caught than an eelpout (p. 143), so for every 44 eelpout there would be only 1.2 rocklings, not 13.4 as in the table. There were some similar complications for the weights of fish, especially because of seasonality, but Table 5.1 gives a reasonable idea.

To obtain insight into possible differences in the quality of the prey for otters, we analysed a number of the fish for their calorific and lipid content. This was done in the Institute of Terrestrial Ecology laboratory at Merlewood, using fish caught in Shetland throughout the seasons. The most striking result was how similar the values for all these fish were. A total of 43 individuals from 10 species were analysed with a bomb-calorimeter, and the highest mean calorific value, in kilojoules per gram wet weight, was 5.01 ($\pm$ 0.17 s.e.; $n = 4$) for butterfish, and the lowest 3.76 ($\pm$ 0.08; $n = 4$) for sea-scorpion. All others were intermediate, for instance eelpout with 4.22 ($\pm$ 0.18; $n = 8$), five-bearded rockling with 3.83 ($\pm$ 0.13; $n = 8$), and saithe with 4.29 ($\pm$ 0.13; $n = 3$) (Nolet and Kruuk 1989). Values for eels (*Anguilla anguilla*), common freshwater as well as occasional marine prey for otters, are higher than those for any

TABLE 5.1 Fish caught with funnel traps in otter study area at Lunna Ness, Shetland 1983–87 (n.d. = no data.)

Species		Per cent of total catch	Median weight (g)
Eelpout	*Zoarces viviparus*	44.1	11.4
Butterfish	*Pholis gunnellus*	7.8	9.6
Five-bearded rockling	*Ciliata mustela*	13.4	26.8
Northern rockling	*Ciliata septentrionalis*	1.3	27.2
Shore rockling	*Gaidropsarus mediterraneus*	2.0	80.6
Three-bearded rockling	*Gaidropsarus vulgaris*	2.5	70.0
Sea scorpion	*Taurulus bubalis*	2.4	15.9
Bullrout	*Myxocephalus scorpius*	1.1	89.3
Sea stickleback	*Spinachia spinachia*	4.6	3.3
Saithe	*Pollachius virens*	12.5	24.9
Pollack	*Pollachius pollachius*	3.4	18.8
Eel	*Anguilla anguilla*	1.0	58.1
Yarrell's blenny	*Chirolophis ascanii*	0.4	n.d.
Two-spotted goby	*Gobiusculus flavescens*	2.1	n.d.
Montague's seasnail	*Liparis montagui*	0.1	n.d.
Topknot	*Zeugopterus punctatus*	–	n.d.
Plaice	*Pleuronectes platessa*	–	n.d.
Cod	*Gadus morhua*	0.4	n.d.
Poor cod	*Trisopterus minutus*	0.2	n.d.
Three-spined stickleback	*Gasterosteus aculeatus*	0.1	n.d.
Lumpsucker	*Cyclopterus lumpus*	–	n.d.
Conger eel	*Conger conger*	–	n.d.
Pipe fish	*Syngnathus acus*	–	n.d.
Dog fish	*Scyliorhinus canicula*	–	n.d.
	Total (*n* = 4141)	100.0%	

other species, at 6.08 kJ g^{-1} (Norman 1963). From this an average prey fish for otters has a calorific content of 4.47 ± 0.06 kJ g^{-1}. We can then calculate that with a mean prey weight per dive of 11.2 g in spring and up to 20.0 g in winter, otters obtain 50–90 kJ per dive, depending on the season.

The fat content of the fish varied somewhat more, and perhaps gives a better measure of the food value of these different prey than the calorific content does (because the calorific content also includes indigestible material). Fat was measured in the traditional Soxhlet apparatus. The highest lipid content was found in three-bearded rock-lings, with 3.35 per cent of wet weight (± 0.12 per cent s.e.), the lowest in sea-scorpion with 0.49 per cent (± 0.07), with others intermediate such as eelpout (1.61 per cent ± 0.17), five-bearded rockling (1.22 per cent ± 0.31), butterfish (2.04 per cent ± 0.28), and saithe (1.17 per cent ± 0.12). In eels lipid content may be as high as 30 per cent (in silver eels before migration), but more usually it is 6–14 per cent, in yellow eels (Boetius and Boetius 1985).

These values also have to be treated with care, because we did not take account of possible seasonal fluctuations in lipid content, and in other species of fish such fluctuations can be considerable (Montevecchi and Piatt 1984). On the present showing, an eel or three-bearded rock-ling is a real prize for an otter, a sea-scorpion a rather miserable catch (and also more difficult to manipulate because of the spines), and the regular prey species such as eelpout and five-bearded rocklings are neatly intermediate in food value. Crabs are the least rewarding prey of all; the actual meat has a low calorific value of about 3.5 kJ g^{-1} (Watt 1991), but even that constitutes only a small fraction of the whole prey animal. A crab is mostly exoskeleton.

Fish activity: time and tide

One of the first things which we wanted to know about fish activity was whether the species in which we were interested were, indeed, more active at night, as the aquarium observations suggested. In that case we would expect to catch more of them overnight in the traps than during the day. Early on in the Shetland study, in April and May of 1984, we set a number of traps, and emptied them at 6:00 h and again at 18:00 h, taking care that we covered periods of high and low tide during both the daylight hours and at night. Figure 5.5 shows the results: many fish and also shore crabs (*Carcinus maenas*) were strikingly more active at night, or at dawn and dusk, which were included in the 'night' period (Kruuk *et al.* 1988). For several species this is confirmed by a Finnish study (Westin and Aneer 1987), as well as by our own aquarium observations.

When one is working along the sea-shore, the state of the tide is of great significance; all animal life appears to be geared to it. We were always aware, almost instinctively, whether the tide was in or out, rising or falling, and when a spring tide was due or a neap tide. The tide af-fected the landscape, the birds, the places where we could walk, our ability to see otters, and the activity of the otters themselves. We asked therefore whether the state of the tide affected the activity of the fish.

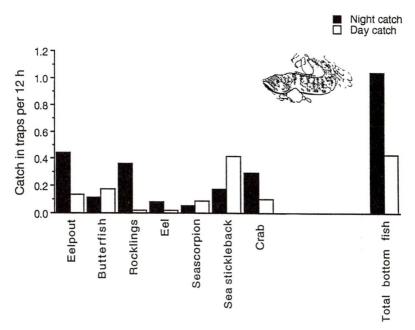

FIG. 5.5 Activity of otter prey in the sea: numbers of fish and crabs caught per trap per 12 hours trapping, at night or during day time (trap collections at 06.00 h and 18.00 h). Differences are statistically significant (Mann-Whitney test) for eelpout and rocklings ($p < 0.001$), for sea stickleback and shore crab ($p < 0.01$) and for all bottom-dwelling fish together ($p < 0.001$; Kruuk *et al.* 1988). $n = 108$ trap days, April–May 1984.

Fortunately, I had an industrious Dutch student, Bart Nolet, who did most of the hard work in studying this question: it had to be done at 3 hour intervals, at all times of day or night.

Traps were left out for 3 hours around high tide, around low tide, and in the periods in between when the water was rising or falling. At the same time, the time of day had to be allowed for so that we got enough tidal observations both at night and during the day. The results were significant: in general bottom-living fish and crabs were most active at high tide (0.3 fish caught per 3-hour period) and least active at low tide (0.17 caught per 3 hours) with catches at falling and rising tides intermediate. Shore crabs, too, were least active during the ebb tide (Kruuk *et al.* 1988). Thus a predator specializing in catching bottom-living fish or crabs during their inactive period would find most available prey at low tide in daytime and this, of course, was exactly what the otters did (p. 162).

There were other variations in our fish catches which were much more difficult to account for, differences which probably had nothing to do with tide, time, or season, and nothing obviously connected with the weather. Some days all our traps would catch many fish of different species and other days nothing: one could expect a certain amount of 'random noise' in the numbers, but it was easy to show that this variation was much more than that. Of course this is an occurrence which every fisherman knows about, and there must be about as many different explanations for these huge fluctuations in the catch as there are fishermen. This variation is something which still irritates me, but which we just had to leave as one of the many unexplained phenomena, one that is just as likely to have affected the otters' fishing success as it did our own.

Variation in fish numbers along the coast

The Shetland coast is remarkable for its differences, with its voes, shallow sloping coasts, or steep cliffs falling into deep water, with a sandy bottom or large rocks, and many different vegetations of algae. One would expect large differences in the fish fauna, and indeed we saw otters emerge with different kinds of prey from dissimilar shores. For this reason we set fish-traps in various places in the study area which I thought represented the different coastal types. The coastal sections differed from each other, for instance in the steepness of the cliffs and shore lines, in the substrate (especially the size of boulders), in their exposure to wave action and in vegetation, and we analysed the extent to which the catch in our traps was associated with these differences. The simplest way to do this was by comparing the mean catches per trap per coastal section throughout the year, with the score of these sections for the different variables, for example for percentage covered by a certain vegetation type.

In the results every species of fish showed its own pattern (Kruuk *et al.* 1988; Fig. 5.6). For example, the eelpout was clearly most common along sheltered coasts (that is there was a negative correlation with 'exposure'), it avoided *Gigartina*, a typical alga of exposed shores, but it was to be found especially amongst the knotted wrack *Ascophyllum* and bladder wrack *Fucus vesiculosus*, the typical vegetation of the sheltered voes. Shore crabs showed the same distribution as eelpout. On the other hand, the five-bearded rockling was found very closely associated with the algae of the really exposed coasts in Shetland, *Gigartina* and the thong-weed *Himanthalia*, and the other rocklings showed even more of an avoidance of sheltered areas with *Ascophyllum* but an association with the indicators of great exposure and with large boulders.

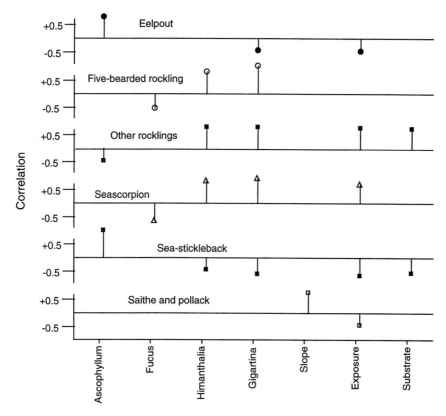

FIG. 5.6 Correlation coefficients for catches in fish traps of otter prey species per section of coast, and different environmental variables of these sections. (Data from Kruuk *et al.* 1988.) Only significant correlations are shown. Variables: percentage of area covered with *Ascophyllum nodosum* (knotted wrack), *Fucus vesiculosis* (bladder wrack), *Himanthalia elongata* (thong weed), and *Gigartina stellata* (carragheen moss) respectively. Slope = mean slope of bottom, exposure = wave exposure index (see text), substrate = substrate size index.

The butterfish and the bullrout were found everywhere, with no particular preference for, or avoidance of, any of the factors we measured, but the sea-scorpion was a fish of the wilder shores, with almost exactly the same habitat preference as that of the five-bearded rockling. The two more mid-water fish, saithe and pollack, showed another pattern again, they were found along steep slopes where the water was deep, with a preference for sheltered areas rather than exposed ones.

Usually we set our traps so that they were 1 or 2 m deep at low tide, but the otters fished over a much wider area, sometimes where the water was well over 10 m deep, or in shallower places. They would spend

different amounts of energy according to the depths at which they were
fishing, and their success rates varied with depth (p. 172). It was obviously important therefore to discover what the differences in fish
numbers were over a range of depths, and in the summer of 1984 we
compared the catches in a line of six trap sites, from 1.5 to 11 m deep, in
one of the voes.

Some of the results of these depth observations are shown in Fig.
5.7. There was quite a variation in the behaviour of the different
species of fish; eelpout and butterfish were more common in shallow
water, but not significantly so, whereas five-bearded rockling, sea-
scorpion, and shore crabs were only found in places that were less than
4 m deep at high tide. The large fish, such as three-bearded and shore
rocklings, occurred more often deep down, as well as some other larger
species; overall, however, there were many more fish in the shallow
strip along the shore. The mean weight per fish was greater the
deeper we set the traps, but the differences were not large enough to

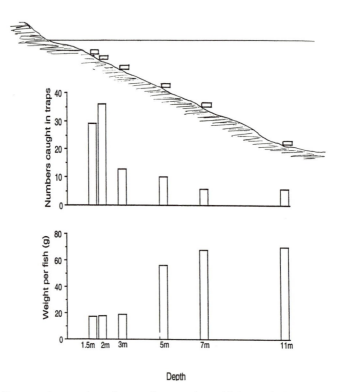

FIG. 5.7 Prey species: total numbers and mean sizes of fish caught in traps at different
depths, over 15 trap days. (Data from Kruuk *et al.* 1988.)

make up for the decline in numbers with depth. In the end the total biomass of fish caught per trap did not change significantly with depth.

These depth transect observations were carried out in summer only, and other times of year could well have shown a different picture with more fish deeper down. There is still a great deal of work to be done on such problems of fish availability.

The implications of these observations for a predator are that in order to exploit eelpout as well as rocklings, sea-scorpions as well as saithe, access is needed to different types of shore, to exposed areas as well as sheltered voes, to steep coasts as well as gently sloping ones. Greater depths should be fished if larger prey are required (for example to provision cubs) and shallow waters, close inshore, are the ones to be exploited for the acquisition of large numbers of small fish.

Seasonal variation in fish numbers, distribution, and weights

Otters need to exploit a range of different species of fish, rather than specialize on just one of them: this became evident when we compared the seasonal availability of each species (Kruuk *et al.* 1988). In Fig. 5.8 the numbers of fish and crabs per trap per day are shown for the various months of the year, for the main prey species; the data are monthly averages over a 4-year period, collected from traps in Shetland's Boatsroom Voe.

From Fig. 5.8 several general conclusions can be drawn. Firstly, every species has its own pattern of seasonality; there was no significant overall correlation between them (Kendall's coefficient of concordance $W = 0.50$, $\chi^2 = 19.4$, d.f. $= 13$, $p < 0.2$). This also meant that our results could not be due to one single, seasonal trap effect, perhaps caused by fluctuations in temperature.

Secondly, every year we found an enormous increase in the total catch in August, due almost entirely to one species, the eelpout, which completely dominated the underwater scene during that month (in every year of our study). There were no clear seasonal patterns for the five-bearded rockling, but butterfish was most abundant in summer, sea-scorpion and shore crabs in autumn, and saithe and pollack in winter. Only the fifteen-spined stickleback, a fish which is a rather miserably small and bony prey for an otter, was clearly more abundant in spring than at any other time.

There were also considerable differences in fish populations between subsequent years, although the seasonal pattern generally remained the same. The data for Fig. 5.8 were collected from 1983 to 1986, but we

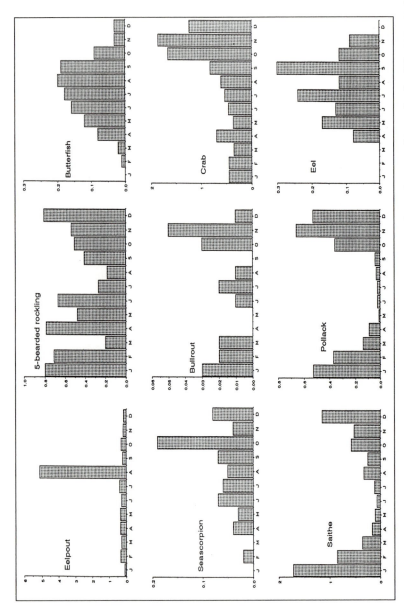

FIG. 5.8 Seasonal catch of some of the most important prey species for otters in Shetland: numbers caught per day for different months, for the whole study area. Note the different scales. The figures are means over 49–219 trap days per month, collected over 4 years. (Data from Kruuk et al. 1988.)

continued to trap in August 1987 and 1988, and there were large differences in the size of the eelpout 'glut' during that month. This variation was important to the otters, especially to the breeding females (p. 231).

To complicate matters, we also found that the seasonality of the various fish species was different along the various sections of coast which we fished, although there were some broad agreements. But comparing the catches for the otters' two main prey species, five-bearded rockling and eelpout, in the different sections of the coast where we had our traps out, there were striking differences. Thus, to efficiently exploit any particular species, an otter would have to fish in different parts of the coast at different times of year. A two-way analysis of variance showed that this complicated interaction between seasonal and area effects was highly significant for all species together, and separately for eelpout, five-bearded rockling, three-bearded rockling, saithe, and pollack (but not for any of the other species we looked at) (Kruuk *et al.* 1988).

There were also seasonal fluctuations in the weights of fish, with median weights of eelpout being about twice as high in August, when they were 16 g, compared with 9 g for January. Rocklings showed a steady 25–30 g throughout the year, and butterfish more than doubled their weight in June (12 g) compared with January (5 g). For further details see Kruuk *et al.* (1988).

Figure 5.8 gives a good impression of seasonal fluctuations in numbers of fish caught in the traps. However, it does not show numbers of different species actually present along the shores, because the species differed in the likelihood that an individual would get caught, in their 'trapability'. In order to correct for this, and in general also to interpret our trap catches in terms of prey density and biomass, we compared some of the catches with what we found in the area immediately around the traps. This was done by detailed searching, hand-netting, snorkeling, and scuba-diving over an area of 20 m^2, where we assumed that we caught all fish present (Kruuk *et al.* 1988).

It was found that an individual eelpout near our traps was 5 times more likely to be caught than a butterfish, and a rockling 11 times more likely than an eelpout. Or, more accurately, in an area where we would catch 10 eelpout per trap per night, there would be 6.7 eelpout per 10 m^2; similarly, 10 butterfish per trap per night corresponded to 31.1 butterfish per 10 m^2, and 10 rocklings per trap per night with only 0.6 rocklings per 10 m^2. This enabled us to calculate correction factors, to translate numbers caught in traps into actual fish densities. Unfortunately we did not get enough data on other species.

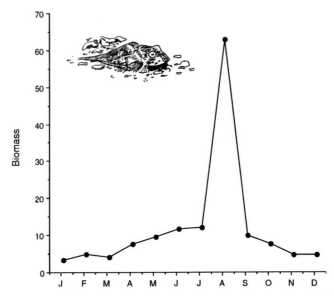

FIG. 5.9 Estimated total biomass of the important prey species for otters in Shetland, in grams per square metre at different times of the year. Data combined for eelpout, rocklings, and butterfish. (From Kruuk *et al.* 1988.)

From the numbers of fish caught, their weights, and the correction factors, I calculated the monthly changes in fish biomass in the otters' habitat (Fig. 5.9). In general, fish biomass was several times higher in summer than in winter or early spring, with an enormous peak, a veritable glut, in August due to the annual eelpout invasion.

Fish removal experiment

We did one small experiment to see what the effects would be when predators such as otters removed a large number of fish from a small patch—the question was, would new fish move in quickly? Clearly, this was a relevant point to settle when we were looking at otters which were repeatedly fishing in the same small patches along the coast.

During a series of very low spring tides, we first removed and counted all the fish which we could find, from under rocks and seaweeds in an area of 20 m², at the water's edge in the Boatsroom Voe, using small hand nets, and taking great care to put stones and weeds back in exactly the same place. We counted 16 fish (mostly butterfish, but also several eelpout and rockling, and some which we could not identify because they escaped), and we managed to catch and remove 11. The next day we did exactly the same thing in the same marked area, and again we counted 16 fish (about the same numbers of each species), catching and removing 14 of them. For the count on the third and last day we found 20 fish, a slight increase due to a larger number of butterfish (Kruuk *et al.* 1988).

It appeared, therefore, that fish caught by a predator in daytime from a small 'patch' are very rapidly replaced, probably during the following night, when the fish were active. Similar results were obtained with butterfish on the Scottish west coast (Koop and Gibson 1991). It is likely that a patch has a certain number of suitable sites for these fish sheltering under the rocks, and vacancies are filled up almost as soon as they arise.

FISH POPULATIONS ELSEWHERE

To my knowledge there is only one study to date in which fish and otter populations have been estimated simultaneously. In our project in rivers and streams in north-east Scotland we studied numbers of fish and their annual productivity, concentrating on brown trout *Salmo trutta* and Atlantic salmon *S. salar*, and comparing that with estimated predation by otters (Kruuk *et al.* 1993). Most of the fish data were obtained by David Carss, using electro-fishing methods.

As an example, Carss found that in the Beltie Burn, a tributary of the River Dee, salmonid biomass varied between 9.2 and 14.4 g m^{-2}, and this population produced 16.1 g m^{-2} year^{-1}, very largely as fish smaller than 15 cm. These estimates appeared to be quite characteristic, and very similar to values found elsewhere in Scotland, England, Denmark, Norway, and North America (Egglishaw 1970; Mortensen 1977; Elliott 1984; Bergheim and Hesthagen 1990; Newman and Waters 1989). Biomass of other fish, mostly eel, was only 0.5–1.6 g m^{-2}. The productivity of the salmonid populations was calculated by comparing fish numbers and sizes in May and September, that is, the beginning and end of the reproductive and growth season, following methods established by Ivlev (1966) and Ricker (1975).

The total salmonid population in our study areas was not evenly distributed through the streams, but it was concentrated in the narrower streams and tributaries, which were relatively much more productive than the wider rivers (Fig. 5.10). Since then Durbin (1993) has found that of the two species we studied, brown trout concentrate in narrower streams whereas salmon tend to stay in wider waters. This distribution of the most important prey species appeared to have a striking effect on the dispersion of otters there (p. 55).

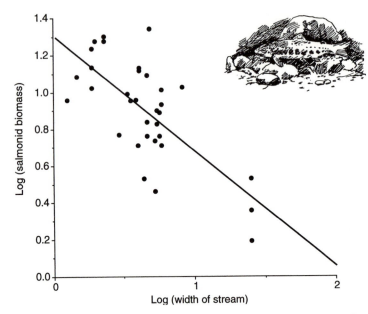

F<small>IG.</small> 5.10 The relation between biomass of otter prey, and the width of streams: salmonids in north-east Scotland, from electro-fishing. Log biomass (grams per square metre), log width (metres): correlation coefficient $r = 0.71$, $n = 34$, $p < 0.001$; $r^2 = 0.50$. Regression $y = 19.01x^{-0.62}$.

OTTERS' SELECTION: DIET AND PREY POPULATIONS

How does variation in fish availability affect otter diet? In other words, which prey (species, size) do otters select from what is available in different seasons, and at various places along the coast and in streams? What general conclusions can one draw from this?

Otters are fish specialists; furthermore, within the range of fish available for instance in their Shetland habitat, they specialize in bottom-living species. Within the category of bottom-living marine animals in shallow water, do otters specialize in any particular prey, or do they take food opportunistically?

When we compare the otters' overall diet (Table 4.1) with what is present along the coast, there are striking differences. For instance the butterfish which, after the eelpout, is the most frequent benthic fish, is not often taken by otters and far less so than eelpout. Butterfish are, if anything, easier to find and catch than eelpout in the same habitat; they are rather sluggish, and when we ourselves were scuba-diving we very often saw butterfish, but for eelpout one had to search. Five-bearded rocklings were eaten throughout the whole year about half as often as were eelpout, although eelpout were almost 30 times more common overall. Somehow, individual rocklings were much more vulnerable to otter predation than eelpout, and eelpout more than butterfish.

Unfortunately we cannot compare the whole Shetland otter diet with the overall trapping results because the traps themselves operated selectively, refusing fish above a given size. This showed especially in winter when otters frequently took fairly large fishes, such as ling, cod, conger eel, and lumpsucker, which we could not catch in funnel traps or only with difficulty. It is not possible, therefore, to calculate a 'selectivity index' for prey species, but there was sufficient evidence to show that otters specialized in eelpout and rocklings when these were available, and took other prey at times and places when the small, bottom-living species were scarce.

Such alternative prey species included saithe and pollack in winter, two species which were always abundant but which only in winter came into the dense algae in shallow inshore water, where they could be caught in our traps and by otters. Saithe and pollack were present throughout the year in huge shoals in open water, and they vividly demonstrated the point that otters need to catch their prey on the sea bottom or inside dense vegetation. When otters could catch saithe and pollack, these were good-sized, profitable prey items.

In spring otters caught many 'other prey', including three-spined and fifteen-spined sticklebacks and butterfish: all small and spiny, and to all intents and purposes rather miserable food. Of these, only the fifteen-spined stickleback was actually more available at that time than at other seasons, the others were eaten presumably just because more profitable species were absent.

A very interesting marine prey was the shore crab, especially because most of the time it appeared to be largely ignored by Shetland otters. There were masses of crabs, and when scuba-diving they could be seen scuttling away everywhere. In our traps they were caught easily, most abundantly in late autumn and sparsely in spring (p. 142). There were observations in our study area of one or two of the young, newly independent otters eating crabs, as did some of the cubs still dependent on their mothers. There was also one very old female with only stumps for canines, Granny, who habitually ate crabs. Apart from those individuals, we noticed that some otters which we had provided with a radio-harness (apparently felt as an encumbrance) ate crabs for the first week or so. From all these observations, I concluded that crabs must be an inferior type of food.

This result was not at all surprising: apart from the risk of being nipped by the crabs' ferocious claws, otters had to spend much time in landing them, and then it was quite a skill to remove the carapace to get at the meat. Otters did not eat the legs or claws, just the soft central contents, a relatively small reward for the lengthy handling time. I noticed that our study area was quite representative of the whole of Shetland as far as these crab observations were concerned; whenever I walked the Shetland shores elsewhere, it was quite obvious from the many spraints everywhere that crabs were eaten only rarely.

In sharp contrast to this, in many places along the Scottish west coast crabs were a very common item in the otter diet, and often they appeared to totally dominate the food, as documented by Jon Watt (1991) on Mull. Here one finds large numbers of spraints, strikingly white and pink, consisting of nothing but crab remains. Despite this, Watt showed that because of the very long handling time, it hardly paid for an otter to catch crabs. Unless they were diving in conditions where they had an extremely low hunting success, otters were better off not taking crabs (in terms of quantity of food per time spent hunting) because they were foregoing chances of catching more lucrative fish prey whilst dealing with the crab (Watt 1991). His elegant calculations suggested that this would be true for Shetland otters even more than for animals in his study area on Mull, because the Shetland fish as prey are generally larger, and more abundant.

Clearly, otters on the Scottish west coast have a much tougher time foraging than do their Shetland counterparts. Watt showed that there were far fewer suitable fish as otter prey in his crab-eating population; also these same otters ate many more butterfish than their Shetland relatives, again a low-quality species. If the Scottish west coast otters caught a dogfish, they would completely eat the small shark, tough skin and all, whereas in Shetland otters would only whip out the large, oily liver and consume that, abandoning the rest: I never saw them eat more of a dogfish than just the liver. The west coast otters also almost always ate whole crabs with the carapace, claws and everything, whereas in Shetland otters would only eat the contents of the thorax and the abdomen from the larger crabs. But if the comparison between west coast and Shetland shows that all this potential food can be used, why do Shetland otters waste such a resource, or ignore it, especially since their food is likely to be limited (p. 235)?

The answer may possibly lie in the seasonal availability of the various foods. For instance, crabs are relatively scarce in Shetland just at the only time of year, spring, when food might be or is at a premium, when much of the otter mortality occurs (Kruuk and Conroy 1991; p. 225), and when there might be a possible advantage in eating crabs. At other times of year there would be no strong incentive to tackle crabs because of the abundance of more profitable prey. Another, or perhaps an addi-tional, explanation could be in Watt's observation that in his population crabs were eaten mostly by young and less efficient otters, which would be less likely to be able to catch suitable alternative prey. Perhaps there was an overall difference in the population compositions of the two areas, with fewer young otters in Shetland at a time when fish was least plentiful.

Apart from the otters' selection of prey by species, selection by size could also be important. The estimated median sizes of fish caught by otters (Kruuk and Moorhouse 1990; p. 107) were larger than those of fish caught by our traps. This was not likely to be caused by the traps being selective for size within this range (Kruuk *et al.* 1988), and the results therefore suggested that (just as in the selection by species) otters took the most profitable prey items throughout most of the year, though perhaps not in spring during low abundance.

This result could be due to a simple difference in the availability of various fish sizes in the otters' likely foraging sites, but as an alternative and more probable explanation otters would actually decide during their foraging along the bottom which fish to take and which to leave. It would be a fascinating study to investigate the factors influencing such decisions.

In Mull on the Scottish west coast, otters ate smaller fish than in Shetland, and prey there was of the same size as that captured from the fish population in fish-traps (Watt 1991). This confirms the suggestion that otters there were exploiting fish populations much more intensively than in Shetland, at least during most of the year.

In freshwater streams in north-east Scotland salmonid fishes (brown trout and Atlantic salmon) dominated the fish fauna (Kruuk *et al.* 1993). They showed a striking age-class distribution with many 1-year olds (3–5 cm in summer), fewer 2-year olds (6–10 cm), and very few older than that (Fig. 5.11). We found that otters ignored the youngest age-class, and took any fish older than that in relation to availability. Similarly, Libois and Rosoux (1989) found in rivers in western France that otters took eels of all size classes as present in the population.

In contrast to this last study, however, David Carss (in preparation) found in otters feeding on eels (Fig. 5.12) in two lochs in north-east Scotland that they selected the larger sizes from the population (com-paring electro-fishing results with eel remains in spraints). The differ-ence in results between the French and the Scottish research could possibly be explained by a difference in habitat exploitation by both eels and otters: in the Scottish lochs there were significant differences in the

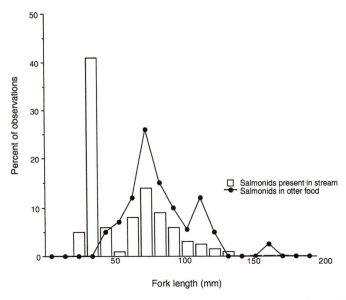

FIG. 5.11 Size of potential and actual prey of otters in streams in north-east Scotland: salmonids in the Beltie (tributary of the River Dee). Columns: potential prey from electro-fishing, June 1989 and 1990, 2718 fish. Curve: actual prey sizes calculated from 40 atlas vertebrae collected from spraints (Feltham and Marquiss 1989). (From Kruuk *et al.* 1988.)

FIG. 5.12 Eel (*Anguilla anguilla*) in fresh water.

size of eels from different places and substrates, and otters could there-fore select prey by fishing in certain sites. In streams and rivers otters are more likely to meet and take eels of all sizes with equal probability. Also in lochs it may be energetically more expensive to land and eat a prey, because of the distance to shore, therefore otters may have to be more selective.

Thus, characterizing the otters' selection of fish: bottom-living species or mid-water fish (in some places also crabs) which can be easily caught when inactive on the bottom, of all sizes over 5 cm. High-calorie prey and larger sizes are preferred and selected for, but in many areas otters take fish of all kinds and sizes.

EFFECTS OF OTTERS ON THEIR PREY

In most of our studies we have been interested especially in the effects of prey species, their density, size, and distribution, on otter diet, behav-iour, numbers, survival, and dispersion. However, there is another side to the interaction, the influence which otter predation could have on the fish populations.

It is often argued that the effect of vertebrate predators on prey popu-lations is small, or at most instrumental in 'regulating' some other major, environmental influence. For instance in East Africa, predators on large ungulates (mostly spotted hyaenas and lions) caused sufficient mortality in a population of non-migratory wildebeest in the Ngorongoro Crater for the wildebeest population to be kept at numbers which the grasslands could support. A neighbouring population of migratory wildebeest in the Serengeti, however, suffered only little predation (the carnivores could not keep up), and the balance between grass productivity and herbivore grass intake was maintained through starvation-mortality of the wildebeest (Schaller 1972; Kruuk 1972; Sinclair 1977). Similarly, predation on grouse in Scotland appears to affect mostly non-territorial birds, a 'surplus' of the population (Jenkins *et al.* 1964). This role of predators appears to be a common one, and we usually assume that the role of predators in a community is not really that dramatic and that they do not actually eliminate prey populations.

However, there is a problem with that assumption. When we study a community of animals as we find it, any elimination of prey species by predators would have taken place long ago. The true influence of carni-vores is demonstrated during or after major perturbations, and a clear example of this was provided by Estes and his colleagues for the sea otter *Enhydra*, along the Pacific coasts of North America (Estes and

Palmisano 1974; Breen *et al.* 1982; Estes and Van Blaricom 1985; Estes 1989). The species was almost wiped out by fur hunters, but conservation legislation became effective and it returned to many areas after an absence of about a century. This was a major success for conservation, but in some areas the fishing industry for shellfish (abalone) went out of business, because of the enormous impact of sea otters on these large molluscs. More dramatic still, their predation on sea-urchins meant that these echinoderms no longer grazed the algae, and this caused the growth of huge kelp beds. In many places sea otters now feed to a large extent on fish, a different prey altogether, and there are very few molluscs left for them to take. The entire community structure was altered on a very large scale, just by the return of the top predator in the system, the sea otter. There are similar cases elsewhere, where introduced predators, for example feral cats on Antarctic bird islands, appear to live in a well-balanced predator/prey system with various species of sea-birds—but where we know from historical records that those predators first wiped out several species (Derenne and Mougin 1976).

These cases should act as a caution when we ask whether otters affect their prey populations. In Shetland it seems unlikely that otters would have a major influence in the shore community; the main argument for this is that the otters only fished in shallow, relatively narrow strips of water, usually less than 3 m deep, and the prey species appeared to go down to greater depths than that. In other words, otters only creamed off the edges of prey populations.

The low potential for a limiting role of predation in the populations of prey species in Shetland was demonstrated most strikingly in the case of the eelpout, the fish which otters took most frequently. It occurred everywhere, also at depths of 20 m and probably more, and enormous numbers migrated into the shallows of the study area in summer, a glut of prey. However, we cannot exclude the possibility that, for example, some other fish species which spawn in shallow water could be hit hard by otter predation. The lumpsucker could be such a case, spawning near the low-tide mark where otters took almost only the males, which look after the brood (Wheeler 1978).

Otters may also have influenced numbers of one of their main prey, the five-bearded rockling: that species was confined to a zone less than 5 m deep. We estimated that otters annually removed over one-quarter of the local population, the 'standing crop' (Kruuk and Moorhouse 1990). Of course, it was not known what proportion this was of the productivity of the population, but otter predation appeared to be quite substantial.

The relations between otters and prey in fresh water are different from those in the sea, in that the whole fish population is accessible to predators. Also, in fresh water it is easier to quantify prey numbers. In streams and rivers in north-east Scotland where we estimated otter numbers (p. 210) and food consumption (p. 123), we also calculated fish density and productivity (Kruuk et al. 1993). For instance in the Beltie Burn, a tributary of the River Dee, otters spent 124.5 nights per hectare per year, consuming 9.6–12.0 g salmonid per m^2 per year. Salmonid biomass was calculated from electro-fishing, and it varied between 9.2 and 14.4g m^{-2} (p. 145), with an annual production of 16.1 g m^{-2} per year. Otters alone, therefore, accounted for 53 to 67 per cent of the annual production of salmonid fish, or 60 to 118 per cent of the mean 'standing crop'. There was a good correlation between salmonid biomass (and hence productivity) and otter utilization of streams ($r = 0.97$, $n = 4, p < 0.05$) (Kruuk et al. 1993).

The proportions of fish productivity taken by otters as predators in the Beltie appeared to be representative for many streams (Kruuk et al. 1993). They are clearly very high, especially when taking into account that other natural fish predators, such as herons (*Ardea cinerea*), goosanders (*Mergus merganser*), mink (*Mustela vison*), and other fish, are also present. Nevertheless, the figures do not imply that otters have a large effect on fish density—they may be taking the part of the fish population which is surplus to the streams' carrying capacity, as in other predation examples presented earlier. The effects of predation on fish density can only be assessed properly by experiment. However, otters clearly play an important role in the interaction between fish populations and their environment.

The above data were collected on the otters' common prey species, in a situation where predator and prey have a long history. As explained earlier, there is at present no way in which one can establish whether otters have caused the demise of previous fish populations, or may have prevented the establishment of other species. Perhaps a change in fish populations in areas where otters have become extinct could provide clues, in the same way as established for the sea otter. However, the causes of otter extinction, such as pollution, may have had direct effects on fish populations as well.

6

Otters fishing: hunting behaviour and strategies

The usual hunting behaviour of otters in the sea consisted of swimming along the surface in daytime, diving in shallow water, and capturing small fish from their hiding places. Swimming speeds are described, as well as diving, handling, and eating of prey. Occasionally otters captured other animals such as aquatic birds, and they caught rabbits in their burrows. Otters were active during the day in Shetland and at night in many other places, because of the activity pattern of their prey species, the majority of which were caught when inactive. In the sea there was also a tidal effect, with otters foraging least during high tide. Success rates appear to be partly predetermined (otters fishing when there is a maximum likelihood of success). Differences in success rates between areas appeared to be related to prey size, and there was little variation in success rates between different depths of water and between seasons, with overall diving success around 27 per cent in the sea.

Otters might dive as deep as 14 m but mostly less than 3 m. Their preference for shallow waters was not explained by differences in fish density or success rates, but could be based on energetic expenditure or thermal insulation. Much of the fishing was done in small patches, where prey was rapidly replaced. Energy expenditure was determined by measuring the oxygen consumption of swimming otters; it was high, and determined to a large extent by water temperature. Rates of fish acquisition required to make up daily energy deficits were estimated, and compared with field observations.

Otter cubs along the Scottish west coast took well over a year to fully develop their fishing skills, increasing their success rates and diving efficiency, and being taught by their mothers. The Shetland otters developed faster, reaching independence within a year.

Foraging behaviour of other species of otter is remarkably similar (except for that of the sea otter), and some differences in detail are described.

INTRODUCTION

Artists' impressions of a fishing otter, the pictures one finds on posters and in popular books, often show the animal in fast pursuit of a shoal of fish. In the wild, however, otters feed on fish which they do not have to chase. Clearly, it is important to understand such details if we are to assess food availability, and the otters' strategies of exploitation, both for purposes of insight in the species' ecology and evolution, and for conservation management. In this chapter I shall discuss observations on the feeding strategies of otters and detailed foraging behaviour, in order to understand how they exploit their resources, how they select prey, and what determines their success and profitability.

These questions are not just an academic exercise. It would be difficult to extrapolate from our Shetland results to predict what otters do elsewhere, in areas where they are less numerous, if one did not understand the mechanism of the relationships which we found. For instance, one may be able to generalize about otter prey preferences, or preferences for particular depths of water, or for fishing at a certain time of day, if we understand why and how in Shetland certain species were taken most often, and what exactly are the costs and benefits of diving in certain places. Such generalizations include the interpretation of the diet and metabolic costs against the background of fish behaviour and availability, which is basic to the understanding of spatial and social organization, and of the 'carrying capacity' of particular waters for otters.

Many of the questions about foraging of otters are also scientifically fascinating on their own, questions about diving behaviour and capabilities, the exploitation of food patches, energetics, and heat loss. Some of the hypotheses I will discuss here are that otters maximize their diving success rate, they fish at depths where they get most prey for least effort, they fish at times of day and tide and in places when and where their diving success is highest. The answer to several of these ideas and questions appeared to be intuitively obvious, at least to me, but my intuition turned out to be at least partly misleading.

FISHING BEHAVIOUR

The following edited extract from my field notes describes the most common type of otter foraging in Shetland.

July 1985, 05.00h. From my hide on the shore of the Boatsroom Voe I watch an otter female walking on the opposite bank, probably just getting active. She spraints, then enters the totally calm sea, swimming in my direction. Just the head shows, occasionally a bit of back; sometimes she swims completely submerged. A heron flies over, two tysties (black guillemots) are fishing nearby, a merganser sits along the shore. When the otter is about thirty metres out from my side of the voe she dives, with very little splash or ripple to show for it: the end of the tail is lifted right out of the water and at the same time she goes down, probably at an angle of some 60°. I can see bubbles rising, the chain trailing her progress deep down: obviously she covers very little bottom, just stays in the same few square metres. I know that it is about four metres deep there at this stage of the tide, and I can picture her, rooting about in the kelp, in the forest of *Laminaria* where she can slip between the stems but where I, as a diver, get completely tangled up if I try to do the same.

Eighteen seconds after I saw her tail-tip slip out of sight, she emerges, almost cork-like, hitting the surface about five metres from where she went down. She was successful: an eelpout about twice the width of her head in length. Head up, chewing, she treads water showing her throat patch for all the world to see, once or twice aiding the processing of the eelpout with one of her paws. Half a minute later the fish has gone, and down she goes again, in the same area. Ten seconds later she comes up with another eelpout of about the same size. The following dive lasts 25 seconds, and she emerges with empty jaws, quietly floating on the surface for seven seconds, then going down again.

The hunting session lasts 24 minutes, during which time she catches 15 fish, almost all eelpout, and all in an area no larger than 20 × 50 metres. She lands, shakes, walks about on the seaweed, and lies down. A bit of a roll, some grooming, then she curls up and sleeps.

In our Shetland study area, this type of foraging was the one used most often; otters swam along the surface, mostly less than 50 m out from the shore, then dived (Fig. 6.1) to the bottom in water usually less than 8 m deep, often going down many times in a small area. The periods they spent under water varied considerably; the longest dive I watched was for 96 s (one of our radio-tagged otters diving in very deep water) but it was quite rare to see dives lasting longer than 50 s. Some published values of mean diving times for otters in the sea are 23.1 s (West Scotland; Kruuk and Hewson 1978), 20.1 s (Shetland; Conroy and Jenkins 1986), 23.3 s (Shetland; Nolet *et al.* 1993), and 22.7 s (Mull, West Scotland; Watt 1991). All these figures are based on large numbers of observations and are strikingly similar.

Prey was consumed on the surface of the water (Figs 6.2, 6.3), except if it was large and/or difficult to handle, such as dogfish (*Scyliorhinus canicula*, Fig. 4.4, p. 99), or if it was to be taken to cubs. Some otters, especially young ones, were more likely to take small prey ashore than others, and if

FIG. 6.1 Deep dive: the otter's tail lifts out of the water.

FIG. 6.2 Most common way of eating prey in Shetland. The fish is an eelpout; note the otter's throat-patch. (Photograph by Thomas Stephan.)

otters foraged close inshore, they would be more likely to land prey than if they were far out. Some small, spiny prey, such as sea scorpion which was difficult to manipulate, were often brought to land (Fig. 4.2, p. 98), and when a crab was caught it was invariably taken ashore (Fig. 4.3, p. 99).

FIG. 6.3 Otter eating prey at the surface of the sea.

Landing prey is time-consuming, and must add a great deal to the energetic cost of foraging (Watt 1991; Nolet *et al.* 1993).

We assumed that when otters were fishing they were actually going down to the bottom, and they did not chase fish in mid-water, because (a) whenever we could actually see what was happening, which involved watching from above in clear water, the otters went right down, (b) the fish they came up with were almost always bottom-dwelling species, and (c) the chain of bubbles suggested that fish were caught on the spot where they were found.

To catch their prey, otters are likely to use both sight and touch (Green 1977), but the sense of touch must have been all-important. When our captive otters dived in an observation tank, one could see the big set of whiskers being turned down and forwards when they were foraging, which provided a maximal contact area for prey disturbed by the ever-questing snout. The fact that mainland otters catch their fish mostly in the depths of night also suggests that eyesight is of very much secondary importance. Once in north-east Scotland, I caught a large, adult male otter in one of our box-traps along the Beltie, a tributary of the River Dee; it was completely blind (white opaque eyes), and in excellent physical condition, demonstrating that its disability had no effect on its foraging success, using tactile stimuli.

One of my students, Addy de Jongh, measured the diving and swimming behaviour of a captive yearling otter by filming the animal, a female named Penny, in a large swimming pool and calculating speeds as described by Videler (1981) for fish. When diving, Penny descended at about 70°, at a speed of 0.62 m s^{-1}; when pursuing a fish, at presumably its optimal swimming speed, she moved at 1.20 m s^{-1}. Penny was not fully grown (she measured about 76 cm in length as opposed to the 1 m of an adult female) and to extrapolate these figures to adult otters, speeds have to be calculated relative to body length (Videler 1981).

One can estimate, therefore, that an adult female otter will swim under water at a speed of about 1.52 m s^{-1}, or 5.5 km h^{-1} (Nolet *et al.* 1993). However, when animals were observed in Shetland searching for food along the bottom of clear water, we calculated (from times taken to cover estimated short distances) that they moved much more slowly than that, their speeds were about 0.26 m s^{-1} or 0.9 km $^{-1}$ (Nolet *et al.* 1993).

The general, most common, foraging pattern we called 'patch fishing' (Kruuk and Moorhouse 1990; Kruuk *et al.* 1990), where otters repeatedly dived and searched in a relatively small area of water. Apart from this main strategy there were a number of variations on the theme. For instance, we recognized 'swim-fishing' which was when otters swam along the surface, then dived to come up again quite far ahead but going in the same direction, and then continued to swim along the coast keeping to the surface, diving again at intervals. This was seen when otters moved between fishing patches, and was especially striking when otters were feeding along steep cliff coasts (often in white water), and were only covering a very narrow strip of sea bottom because of the steep incline. Another foraging strategy we termed 'kelping', which was when otters were feeding in exposed seaweeds at low tide (Figs 6.4, 6.5, 6.6). Presumably they did very much the same thing then as during their usual feeding dives in deeper water, but it looked different to us because the kelp showed, as did the otter, wriggling under and through the dense mat of fronds.

In freshwater lakes otters' fishing behaviour is very similar to that described for the sea, with diving times of the same length for given depths (see below). Underwater times in the relatively shallow waters of rivers and streams are correspondingly short, and the otters were almost always on the move, with very little 'patch fishing'.

In all freshwater habitats, prey is landed much more often than in the sea (where it is usually eaten by the otter whilst floating on the surface), and there are several reasons for this. For instance, in lakes otters often took eels which are a difficult prey to manipulate, even apart from their relatively large size, and they have to be dealt with ashore. Otters fishing in rivers and streams catch most of their food

FIG. 6.4 Foraging ('kelping') amongst exposed knotted wrack (*Ascophyllum nodosum*).

FIG. 6.5 Foraging ('kelping') in shallow water in serrated wrack (*Fucus serratus*).

FIG. 6.6 Fishing at low tide in bladder wrack (*Fucus vesiculosis*).

close to the bank, and they have the current to contend with when trying to remain stationary. This, too, made it easier for the animals to eat on land. Carss *et al.* (1990) described otter predation on large salmon in Scotland, where fish were taken predominantly in sections of river where the water was very shallow with riffles. Sometimes the salmon could be eaten on the riffles where they were caught (Fig. 4.22, p. 121), splashing about in only a few centimetres of water. Most of the salmon caught (81 per cent in a sample of 63) were male, presumably because males spent more time travelling up and down the river and were a comparatively easy and only partly submerged prey on the riffles. During a period of 9 days' intensive observation, one large male otter killed one salmon each night, of which he consumed approximately 1 kg per fish, and this appeared to be fairly typical.

In Shetland we also saw otters using some of their more rare and specialized hunting methods on prey other than fish. We made observations of them catching birds and rabbits, behaviour which was both conspicuous and curious, but in general it did not contribute much to the diet. An otter would dive and then come up 30 m away or more, right underneath a swimming merganser, or a shag, or a fulmar—which would explode out of reach if it were lucky, or be dragged below the surface if it were not. Once we saw a fulmar taken on the cliff by a large male otter, which more or less accidentally cornered the bird in its nesting hole. Birds are similarly taken as occasional, one-off prey by sea otters *Enhydra* (Riedman and Estes 1988, 1990) and by the American river otter when it lives in the sea (Riedman and Estes 1990), both species using the same method of capture as did our otters. On the Scottish mainland I have documented several cases of otters taking domestic fowl such as geese and hens (including three of my own ducks at home, where I could follow the tracks of the predator in the snow).

Rabbits were always caught underground by otters entering rabbit burrows. One Shetland otter, initially with a radio-collar, was named Miss Rabbit by us, and she was a real specialist in this, with rabbits featuring conspicuously in her spraints for at least a year. She would enter rabbit holes close to the shore of fresh water lochs, emerging with a rabbit which she would then consume a short distance from the warren. I assume that rabbits are caught similarly by otters living in freshwater areas, because some of our otters with radio-transmitters occasionally slept in well-used rabbit warrens, and their food also included rabbits (Kruuk *et al.* 1993).

In freshwater areas in winter and in spring, otters take many frogs and toads, and clearly they are seasonally very important prey (Weber 1990). Frogs are caught in winter at the bottom of muddy stretches of streams, and in spring in marshes and shallow ponds away from the otters' main river and lake habitats. The unpalatable egg-mass of a female frog caught before spawning is tell-tale evidence left on the bank. Toads are skinned before being eaten, and the skin and much of the skeleton is conspicuously discarded (Weber 1990).

Observations such as the above, where otters catch relatively large or spectacular prey, get much attention, but neverthless the vast majority of otter predation attempts consist of diving for relatively small fish, both in the sea and in fresh water.

WHEN TO FEED: TIME OF DAY AND STATE OF TIDE

During the short days of winter in Shetland, otters showed one clear peak of activity just before midday, but in summer their swimming and foraging was spread throughout the daylight hours, with one main peak early in the morning and a lower one in the late afternoon (Fig. 6.7). We did not come to this conclusion because we only watched otters during the day, when they were easy to see: we were also able to follow the animals fitted with radio-transmitters throughout the whole 24 hours, and they were only active in daytime. Moreover, we ourselves were often out and about in the morning before the otters were, and we would actually see them starting out. Thus, the animals were clearly diurnal,

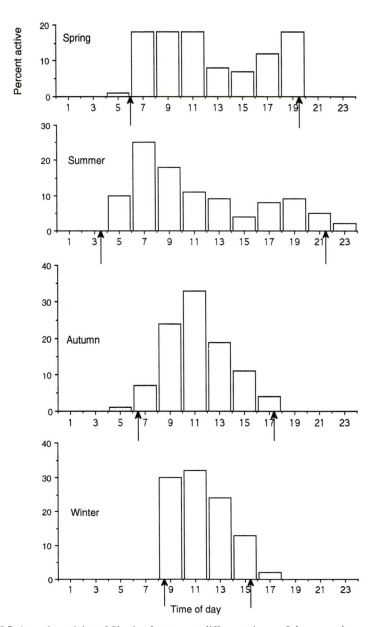

FIG. 6.7 Aquatic activity of Shetland otters at different times of day over the seasons. Based on numbers of dives observed: n = 4781 (spring), 3063 (summer), 3466 (autumn), 2003 (winter). Arrows indicate sunrise and sunset for mid-points of the season. (From Kruuk and Moorhouse 1990.)

their feeding was a daytime activity in Shetland (Kruuk and Moorhouse 1990), as it was on the Scottish west coast (Watt 1991). This is in contrast to otters' general nocturnal nature elsewhere (Chanin 1985; Mason and Macdonald 1986; and our own data).

The most likely explanation for this variation in activity pattern lies in the habits of the otters' main prey species, the bottom-dwelling fish (Chapter 5; Westin and Aneer 1987; Kruuk *et al.* 1988). In Shetland these fish are active at night, swimming fast and often into mid-water, but hiding under stones and weeds during the day. For me, acting as a predator on fish in the aquarium or when diving, it was infinitely easier to touch a 'prey' in daytime than at night. I found my quarry just by rummaging between stone, rocks, leaves, and litter, and it would be the same for an otter.

This explanation would hold true for eelpout, rocklings, and crabs, but some other prey were more diurnal. For instance, sea-scorpions, bullrouts, and butterfish appeared to be active at any time: but these fish were slow, slow in swimming and slow in reacting to danger. They relied on their camouflage and were easy to catch even when they were active in daytime.

In freshwater areas of Scotland, otter food, caught at night, consisted primarily of salmonid fish and eel (Jenkins and Harper 1980; Carss *et al.* 1990; Kruuk *et al.* 1993). Eels are active at night and not during the day (Moriarty 1978); however, even at night they spend long periods lying still either on top of the substrate or partly buried in it. Moreover, in areas where otters feed predominantly on eels, in lakes for example, they often do so in daytime (Conroy and Jenkins 1986; personal observation), just like otters foraging in the sea.

Salmonids and many other freshwater fish are often inactive during the middle of the night, lying on the bottom (Westin and Aneer 1987), and it is then that they would probably be most easily caught by an otter. However, Heggenes *et al.* (1993) and Fraser *et al.* (1993) found that in winter at temperatures below 10 °C, salmonids are more active at night, at least in some northern rivers, and their behavioural periodicity is not as simple as was thought at first. Large male salmon move up and down rivers during the early part of the night, when otters can catch them easily on the shallow riffle areas (Carss *et al.* 1990).

In the sea, the state of the tide also has an important effect on the activity of almost all coastal life. Along the Shetland coasts, otters were significantly less likely to be fishing during the 3 hours around high tide, and they were more active during the same periods at falling, low, or rising tides. Over all our observations of otters swimming, 21 per cent were around high tide, 27 per cent at falling tide, 25 per cent at low tide, and 27 per cent at a rising tide ($n = 1170$, $\chi^2 = 11.83$, $p < 0.001$) (Kruuk and Moorhouse 1990). In reality the preference of otters for feeding at low tide was emphasized considerably more than shown in our data, because it was much more difficult to find otters foraging in the exposed algae at low water, with all the fronds waving about on the surface. This observational bias was evident once we had spotted an otter, because then the number of dives per observation was greater at low tide than at other times (Kruuk and Moorhouse 1990). Again, these figures can be explained by the hypothesis that otters were active when their prey was not, because almost all the fish on which otters prey were more active at high tide (p. 137; Kruuk *et al.* 1988).

FEEDING SUCCESS

Foraging success appears to be easily measured in otters when they are fishing in open water in daytime, as one counts successful and unsuccessful dives. It is important that such a measure of success is established, because we need to compare the efforts of different animals and populations, in different habitats, and so on. However, there are snags and these problems have been discussed at some length (Kruuk *et al.* 1990; Ostfeld 1991). In our Shetland studies we measured success rates during thousands of observations (overall 26.9 per cent of 13 313 dives were successful) (Kruuk and Moorhouse 1990), and we first became suspicious of what we could conclude from these estimates when we noticed how small the differences in success rates were under various types of conditions.

For instance, we compared a set of observations in Shetland of otters which spent most of their time foraging in small sections of coast which we called 'feeding patches'. During this 'patch fishing', which was clearly their preferred mode of fishing (see below), their success was 25 per cent (that is, 25 per cent of 2796 observations were successful). In neighbouring sections of coast, though, on either side of these patches, and where they dived less often, this figure was 23 per cent (485 observations), and in sections beyond those it was 25 per cent (555 observations). None of these differences in success rate were significant, which was surprising because the otters showed such a large preference for the small feeding patches. Similarly, when comparing success rates for spring, summer, autumn, and winter for all otters over the whole Shetland study, we found respectively 24 per cent (number of dives $n = 4781$), 28 per cent ($n = 3063$), 25 per cent ($n = 3466$), and 33 per cent ($n = 2003$), differences which were statistically significant because of our large sample sizes ($\chi^2 = 58.3$, $p < 0.001$). Nevertheless these differences were very small, where we had expected something much more spectacular (Kruuk and Moorhouse 1990).

It seems likely that interpretation of such figures is made difficult for two reasons:

1. Dives differ in length of time, depth, degree of cooling experienced by the animal, and in other ways, so they are not a good unit of effort.

2. A dive is the result of a decision taken by the otter, on the surface, which may be based on an assessment by the animal of its chances of success.

We may therefore be measuring an otter's ability of risk assessment, rather than the risk itself. For instance, an otter might have a strategy in a certain area (with a given depth and mean prey size) which dictates that it dives only if there is at least a one-in-three chance of catching a fish there. Such assessment would be based on the animal's previous knowledge of the sites.

We must then be careful about how to interpret success rates; at least in some comparisons they will tell us very little about the actual effort on the part of the otter, but in other situations they might be useful. Ideally, one should collect data on the total energetic cost of catching and handling a prey, and the reward to the animal in terms of calories gained. Ostfeld (1991) argued that success rates of fishing sea otter and marine otters (*Lutra felina*) show important differences, for example between animals specializing in eating invertebrates and those catching fish. The differences which he reported were greater than the ones we found, for instance sea otters in the Aleutian Islands emerged successfully in over 90 per cent of dives (Estes *et al.* 1981), but on a Californian coast in only 35 per cent (Ostfeld 1982). However, in themselves such variation means little unless one knows more about the nature of the dive (length of time and depth), and the kind of prey (sea urchins or fish, of different weights).

Some of the variation in diving success which we saw in the Shetland otters, although small, was important when considered together with other information (Chapter 4). For instance, in spring success was lower than at other times (24 per cent compared with 33 per cent in winter), estimated mean weight of prey was lower, and the mean weight of prey caught per dive was only 11 g then compared with 20 g in winter. Also in spring, many of the prey items had low calorific values, for example crabs, sea-scorpions, and sticklebacks (Kruuk and Moorhouse 1990).

This variation was statistically highly significant (success rates by season $\chi^2 = 11.7$, $p < 0.01$), and very suggestive of seasonal difficulties experienced by otters. We did not, however, measure how much time otters had to spend in the water to catch a prey, or the seasonal differences in the depths at which prey was caught. Such measurements still need to be made, and they are crucial to the understanding of seasonal stress (see p. 180).

There are large differences in the success of otters in different areas. For instance, Conroy and Jenkins (1986) compared diving success rates of otters in a Scottish freshwater loch and along the Shetland coasts (7 and 27 per cent, respectively). To interpret such figures, however, one has to take into account that in the sea otters took small fish, 20–30 g from considerable depths (usually 1–8 m), and in fresh water the much larger eels (usually 50–200 g) from waters 1–2 m deep. As in our own study, more information is needed here about the length of time spent by otters in the water.

Also the variation in success rates of otters foraging along different types of coast in our Shetland study area was considerable (Kruuk and Moorhouse 1990). Along the most exposed shores and the steep cliffs the sucess rate per dive was only 12 per cent, but in the larger voes and along more sheltered coasts it was as high as 32, 25, and 34 per cent. The associated weights of prey caught in the more difficult exposed areas, however, compensated for this: the weight of prey caught per average dive (including the unsuccessful ones) along the most exposed coast was 19 g, but it was significantly lower in the two main voes, only 9 g and 13 g. In other words, otters along exposed coasts had fewer successful dives, but came up with much larger prey. One very small voe, Feorwick, did not fit into this pattern; otters there had a low success rate, only 16 per cent, and small prey (of a mean weight of 5 g per dive). Feorwick, though, was unusually shallow (rarely more than 1.5 m deep), so a dive there was short and cheap in energetic terms. All this suggested strongly that there were predetermined chances of success for foraging sessions in different areas, perhaps by otters 'knowing' the particular underwater sites with the stones and holes where fish could be found, and where there was an established expectation of success.

Overall seasonal differences in success rates might be small, but there was substantial variation between subsequent years (Nolet et al. 1993). These authors noted that in Lunna, the main study area in Shetland, success at 1, 2 and 3 m was 29, 30, and 38 per cent in 1983, whereas in 1984 it was 15, 13, and 17 per cent, and in 1985 it was 23, 20, and 21 per cent, all in the same sites. Thus success rates can be almost twice as high in one year compared with the next one; the causes of this were entirely unknown, as the annual differences in fish densities (sampled with our traps; see Chapter 5), did not match the otters' success.

On p. 180 I shall argue that at least for some purposes it is more useful to estimate success as reward per time spent in water, rather than per dive.

PATCH FISHING, PREY SITES, AND PREY REPLACEMENT

A phenomenon which has attracted a great deal of attention in many studies of foraging behaviour of animals is that they often feed for considerable periods in relatively small areas. This observation is of great importance, amongst others, in hypotheses on feeding efficiency ('optimal foraging'; see, for example, Krebs (1978)), and in explanations of animal dispersion, territory size, and group size (see, for example, Bradbury and Vehrencamp (1976) and, for carnivores, Mills (1982), Macdonald (1983), Kruuk and Macdonald (1985), Carr and Macdonald (1986), and Bacon *et al.* (1991)). As mentioned already in the general description above (p. 159), otters in Shetland also showed a strong tendency to fish in certain small 'patches', and this is relevant when we attempt to understand habitat preferences, spatial organization, and the factors underlying their foraging efficiency (Kruuk *et al.* 1990).

To quantify the 'patchiness' of the otters' foraging effort, we divided each of the 350 m observation sections of the coast of our main Shetland study area (Fig. 2.3, p. 33) into four subsections, and we plotted the number of times we saw otters dive in each of 178 subsections. Figure 6.8 shows that (a) certain sections, when visited, were fished with many more dives than one would have expected if the otters had randomized their movements, and other sections received far fewer dives, and (b) this same pattern was seen repeatedly, not just during one trip but over many months and probably years. When an otter was fishing along the coast, its median number of dives per subsection was 7.0, but it might dive as often as 61 times in one site.

We arbitrarily defined a section as a 'feeding patch' if the mean number of dives per observation there, over the whole study period, was at least 15, that is at least twice the average number of dives per section. Using this definition, 17 of the 178 subsections (10 per cent) were classified as feeding patches, in which we saw 2796 of the total of 7891 dives (35 per cent).

We then tried to describe how these feeding patches differed from the rest of the coast (Kruuk *et al.* 1990). They were not significantly

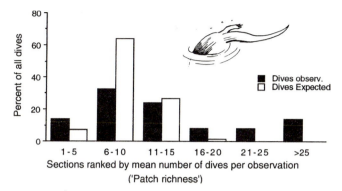

FIG. 6.8 Concentration of otter dives in sections of Shetland coast. Observed proportion of all dives in 75 m sections compared with calculated random expectation. The sections are ranked in order of numbers of dives per observation. Otters dive more often in the 'feeding patches' than expected. (After Kruuk *et al.* 1990.)

different from elsewhere in their proximity to a holt, or to fresh water, and there was nothing else along the coast that appeared to be different about them. We sampled the fish population in the feeding patches with our fish traps, and we compared the results with control patches (of the same depth but 100 m away). There was hardly any difference, in either species composition or biomass.

Next we looked at the otters' fishing success (rates of successful dives) in patches; the mean success in a patch was 25 per cent with 23 per cent in the neighbouring section, and then 25 per cent again in the next section along—there were therefore no substantial differences. Overall, there was no correlation between the number of dives in a section and the success rate there (Spearman rank-order correlation coefficient: $r_s = -0.001$). Finally, we tested to see if perhaps otters in a 'run' of dives on a patch showed a decreasing success rate, following the classical pattern of initial high success, and then subsequent decrease, until the patch became so unprofitable that it would pay to move on again (Krebs 1978). Such a pattern might have been obscured when scoring an overall success rate per patch. In a sample of 20 long runs of dives in one place (total 749 dives), the mean success was 31 per cent for each of the first 10 dives, 30.5 per cent for each of the last 10 dives, and 31 per cent for each of the last five dives. Basically, therefore, there were no differences in success rates during each period of exploitation of a patch.

During these observations, however, we noticed that there was something peculiar about the algal vegetation of the sections with feeding patches, mostly in kelp (*Laminaria* spp.) These sites had more openings in the otherwise solid canopy of fronds and more places where we could actually see the bottom from above; they were sandy places with clear boundaries. The difference with neighbouring control sections was statistically highly significant (Wilcoxon test, $p < 0.005$).

That these openings in the kelp beds could be very important we noticed ourselves when we were scuba-diving, attempting to get at the sea bottom. Outside the open sandy places the kelp is almost impenetrable, and we had to fight our way through the fronds and stems. But in the open spots we could get straight down, then move horizontally in between kelp stems. Otters probably experienced similar problems.

It was likely, therefore, that the main difference between the highly favoured feeding patches and the rest of the coast, was that they afforded relatively easy access for the otters to particular sites (Kruuk *et al.* 1990). This was, therefore, a matter of prey availability rather than prey numbers or biomass. One of the interesting aspects of these observations was the similarity in success rates between patches and elsewhere. This suggested that what we, from the shore, called a distinct feeding patch was no more than an assembly of prey sites, sites in each of which otters could pull fish from under a stone, and that were apparently well known to individual exploiting predators. The likely, and simple, explanation for the similarity in success rate is that otters visit numbers of known prey sites, which are sometimes far apart and sometimes close together (a patch), determined by access through the kelp.

In Scottish freshwater lochs similar feeding patches occur (unpublished observations), small areas close to or somewhere far from the shore where otters dive scores of times and catch eels, year after year. I have not been able to associate these areas with habitat features, but it seems likely that they will be characterized by particular substrates.

Interestingly, otters can harvest these prey sites again and again, given a reasonable time interval. We emulated this process in Shetland in the fish removal experiment described on p. 144. This demonstrated that an area, of the size of a small otter fishing patch, with eelpout, rocklings, and butterfish, can be harvested totally by a fish predator. Yet within 24 hours all the vacancies will be filled again with other potential prey individuals which shelter there.

I believe that this is a key observation for the understanding of otter foraging and spatial organization; food resource exploitation is based on a patchy availability, which is to a large extent both recurrent and predictable to the exploiting predator, if no competing individuals utilize the same patches.

DEPTH OF DIVES

The deepest dive recorded in our Shetland study was 14 m (when otters surfaced with bottom-dwelling prey in waters of known depth), and once in the Western Isles of Scotland, an otter was caught and drowned in a lobster pot set at 15 m deep (Twelves 1983). However, the vast majority of dives occurred in shallow water: for instance, in Shetland, during 1983 to 1985, 54 per cent of 3558 dives were in water that was less than 2 m deep and 98 per cent were in less than 8 m (Nolet *et al.* 1993). When we compared dives in the strip of water within 100 m of the shore, there was a highly significant preference for shallow places (Fig. 6.9).

The reason for this preference was not immediately obvious, and we tested several different hypotheses to explain it (Kruuk *et al.* 1985; Nolet *et al.* 1993). The three main hypotheses were that otters used shallow waters more because:

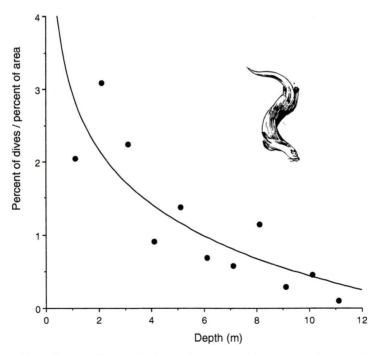

Fig. 6.9 Otter dives at different depths: preference for shallow waters. The graph shows the proportion of 1008 dives at a given depth in 1983, divided by the proportion of area of sea bed of that depth which was available in the 100 m strip used by the otters. $r^2 = 0.70$. Lunna, Shetland.

(1)   there were more fish there, or;

(2)   it was energetically more efficient to dive there (either because
        more underwater time could be spent on searching instead of on
        travelling to and from the bottom, when the area was shallow; or
        because the greater depths only occurred further offshore, which
        therefore involved more surface travel; or because breathing was
        more efficient during shallow dives), or;

(3)   the animals would lose less body heat.

Hypothesis (1): In our fish-trapping observations (p. 140; Kruuk *et al.*
1988), we did indeed find more fish in shallow water, but they were
smaller, and in terms of prey biomass (weight per unit area) there were
no significant differences with depth. Also, the otters' fishing success
(that is, the weight of prey caught per unit of time spent diving) was not
significantly higher in shallow water (Kruuk *et al.* 1985). Prey was larger
when caught in deeper water, but the handling time of prey (the time
spent in transporting and eating) increased proportionally with the size
of it, and overall there was no difference in 'profitability' of dives at different depths (Nolet *et al.* 1993). We therefore rejected this hypothesis:
otters did not prefer the shallows because these were the richer feeding
grounds.

Hypothesis (2): On the face of it, there would appear to be less effort
involved in shallow diving, because of the shorter distance to the
surface. On the other hand, however, once at the bottom an otter will
have to work less hard to stay there when the water is deep, because of
its positive buoyancy which is determined by air in the pelt and the
lungs, and the volume of that air will be more compressed with depth. It
was possible that otters compensated for this by taking in less air when
making a shallow dive than for a deep one, and this could be the reason
why, in shallow water, otters do not remain under water for as long as
they do in deep dives. There was a direct linear relationship between
dive depths and the duration of single dives (Fig. 6.10). Nolet and Kruuk
(1989) and Nolet *et al.* (1993) found that otters' fishing bouts in
Shetland lasted longer if they included fewer deep dives, and the subsequent recovery on land from the hunt was shorter. This suggests that
fishing at greater depths is energetically more demanding.

One more variable which has to be considered is breathing efficiency:
in some diving animals, the longer they have been under water, the
more the proportion of total time they have to spend on recovery
('catching their breath') increases (Kramer 1988). This could make it
advantageous to dive for short periods only, and that could imply diving

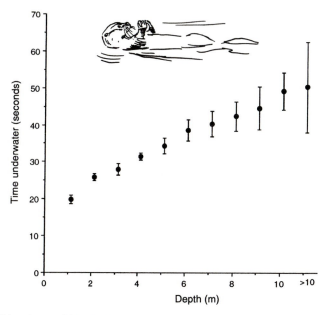

FIG. 6.10 Dive times of Shetland otters at different estimated depths, 1983. Means ± standard error; total number of dives = 1181. (Data from Nolet *et al.* 1993.)

preferentially in shallow waters. However, in otters the recovery time on the surface, after a dive, is directly proportional to the diving time (Nolet *et al.* 1993) (Fig. 6.11), so this in itself could not explain a preference for short dives. Possibly, otters had to breathe harder (deeper and/or faster) after a long dive to affect a quick recovery, but this we were unable to measure.

In freshwater lakes, incidentally, otters frequently come to the surface for only a fraction of a second before diving again, just long enough to take a quick gulp of air, and they can keep this up for extended periods (unpublished observations). This occurred especially in winter and early spring, at low water temperatures, and it was not seen in otters diving in the sea, where even in winter water temperatures are higher. This may be a means to shorten total time in water without reducing prey-searching time, and it will be further discussed on p. 180.

It was possible, therefore, that in the sea shallow dives were energetically less demanding than deep ones, for a similar return of prey, but we were unable to measure this. The only clue lay in the fact that the duration of hunting bouts was shorter when diving in deeper water.

Hypothesis (3): The 'heat loss hypothesis' suggested that the thermal insulation of otters would be less in deeper water, because of greater

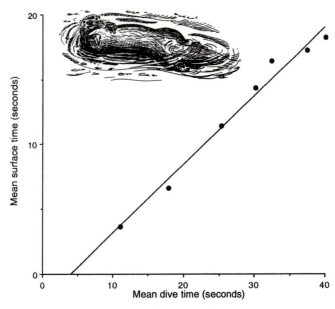

FIG. 6.11 Surface times, compared with dive times of Shetland otters. Figures are means for different depths, total number of observations = 1039. Regression $y = -2.2 + 0.53x$; $r^2 = 0.98$.

compression of the air layer in the pelt. The importance of this insulation to otters has been discussed elsewhere (Tarasoff 1974; Kruuk and Balharry 1990; see also Chapter 7). It was found that otters groomed for longer if they had been diving more at greater depths. However, heat loss could not be measured directly when otters were in either shallow or deep water, so we have no direct evidence (Nolet and Kruuk 1989).

Summarizing observations under the three hypotheses: it was not likely that the otters' strong preference for fishing in shallow water could be explained by greater fishing success. However, it was possible that shallow-water fishing was either energetically less demanding, or that it involved a smaller loss of body heat through more effective thermo-insulation.

ENERGETICS OF SWIMMING AND DIVING

Because otters obtain almost all their food from the sea, lakes or streams, the energetic costs associated with their foraging are very largely the costs of being in the water, of swimming and diving. There were several early indications that these costs are high; for instance, we noticed the trouble the animals took over problems with thermo-insulation (Chapter 7). Thermo-conductivity is about 23 times higher in water than in air (Schmidt-Nielsen 1983). Just seeing the animals swimming in the north of Scotland, for long periods in ice-cold water, without the protection of a thick layer of blubber, makes one shiver.

Several authors have remarked on the high basal metabolism of otters in relation to their aquatic life style (review in Estes 1989), and it is clear that the use of watery habitat by the animals has an important bearing on energy consumption. This, in turn, has implications for the amount of food they have to acquire. Questions arise, therefore, on the costs of swimming and diving under different conditions, on how these costs compare with actual food intake of the animals, and whether these costs affect the exploitation of prey populations.

In this section I will discuss the effects of water temperature, as well as the behaviour of the otters and the effects of other variables, on energy consumption. Energy consumption is normally measured as the amount of oxygen which an animal uses. We had several captive otters, and we estimated their oxygen intake when swimming and diving, which could then be compared with the calorific content of food actually acquired during foraging periods in different areas. The effects of water temperature on body temperature will be discussed on p. 186.

The costs of swimming in cold water have been measured previously in sea otters (Morrison *et al.* 1974), platypus (Grant and Dawson 1978), muskrat (MacArthur 1984), mink (Williams 1986), beaver (MacArthur and Dyck 1990), and others. In some species there is very little variation in heat production in waters of different temperature, for example the Australian water rat (Dawson and Fanning 1981), but in others there is a large and linear increase of metabolic rate in waters of decreasing temperature, for example in beavers (MacArthur and Dyck 1990). If such an increase occurs in otters, it would be an important variable to consider in the cost of foraging.

The oxygen uptake of otters, active in water, had to be studied on captive animals (Kruuk *et al.*, 1994*b*), as it was not possible to do this with wild otters. In the Institute of Terrestrial Ecology in Banchory, Scotland, we kept four otters in two large enclosures, with a big swimming pool and underwater observation windows. In the pool we made a metabolism chamber, which the otters could enter at will, and in which they could breathe air from under a Perspex dome. They could sit on a shelf in the chamber (only their head above water), or they could swim about; the size of the chamber was $1.3 \times 1.6 \times 1$ m. The otters were used to the chamber and frequently played in it after entering through the underwater door.

About once per day, when we wanted to measure the oxygen uptake in one of the otters, we closed the door behind it as soon as it entered the chamber, and kept it inside for 20 minutes, whilst we measured the oxygen flow under the Perspex dome (Fig. 6.12). At the same time the animal's behaviour was recorded; occasionally, an otter was obviously distressed about being confined, and then we terminated the test before the 20 minutes were up.

This experiment was repeated many times on days between August and December, so we had a large range of water temperatures (between 2.1 and 16.6 °C). Each test was split into 5-minute periods, of which we ignored the first one (because the system needed about 3.5 minutes to settle down to a steady flow of oxygen). Altogether we measured 208 5-minute periods, in which otters were sitting on the bench for 46 per cent of the time, quietly swimming and diving for 27 per cent, and swimming fast or vigorously scrabbling at the underwater door also for 27 per cent of the time. This was very much dependent on water temperature (T_w), with positive correlations (r) between T_w and bench-sitting ($r = 0.19$) and between T_w and swimming quietly ($r = 0.39$), but negative between T_w and scrabbling ($r = -0.36$). These results were all

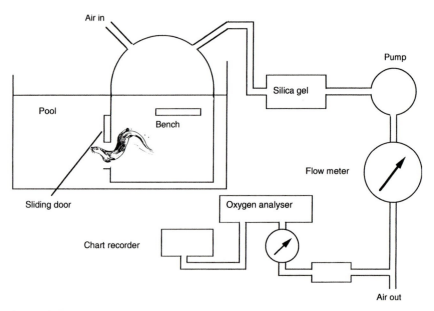

FIG. 6.12 Diagram of experimental set-up to measure oxygen consumption of otters swimming, or sitting still in water. The pool measured $5.0 \times 2.3 \times 1.2$ m, the metabolism chamber $1.6 \times 1.3 \times 1.0$ m, with a Perspex dome roof forming an above-water volume of 120 litres of air. (After Kruuk *et al.* 1994a.)

highly significant, showing that the colder the water, the more vigorous the animals' behaviour and attempts to leave the water, and during warmer days they were more likely to sit on the bench.

Figure 6.13 shows the relation between water temperature and the otters' oxygen consumption. There was no doubt that the animals expended much more energy when swimming in the cold. When we analysed the relationships with stepwise multiple regressions we found that 55 per cent of the variation in oxygen consumption of the otters could be explained by differences in T_w (a correlation of $r = 0.73$), independent of behaviour. Additionally, differences in behaviour of the otters explained a further 14 per cent, so T_w and behaviour alone accounted for over 69 per cent of the observed variation in oxygen consumption. We introduced other variables into the equation, such as individual differences between the animals, the otters' body weight, the time that animals had been in the water, but they had very little effect. Interestingly, if one extrapolates the graph in Fig. 6.13 to the point where oxygen consumption would be zero, one finds that, as expected,

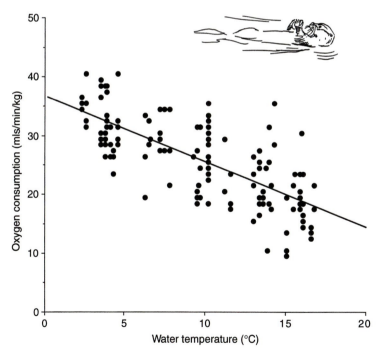

FIG. 6.13 Oxygen consumption by three captive otters (in ml of oxygen per minute per kg of body mass), in water of different temperatures, irrespective of their behaviour.

this occurs at a water temperature $T_w = 37.6$ °C, which is approximately body temperature; this gives confidence in our measurements (Kruuk *et al.* 1994*b*).

Expressed in terms of energy expenditure E (in W per kg body weight), and with 'average' swimming behaviour

$$E = 12.1 - 0.40T_w.$$

This approximates the cost to an otter of catching a fish, for a given time in the water, and it can be used to assess foraging profitability.

In estimating the costs of swimming and foraging in water, we have to make a number of assumptions. Some of these we cannot really test, but at least these assumptions provide us with a start. For instance, the behaviour of our captive animals will have to be taken as representative for an otter swimming and diving. Clearly there are differences, but in our observations in the metabolism tank there was no significant variation in oxygen consumption when animals were sitting still on the bench under water compared with when they were swimming and diving quietly, and in general the variation in the otters' behaviour was less important than the variation in water temperature. On the other hand it was suggested from observations in Shetland that a possible reason why otters dive more in shallow water is that they lose less heat there (fur insulation being more effective in the shallows) (Nolet *et al.* 1993). Thus, at least some of the foraging in the wild in deeper water will be more expensive than we estimated in captivity. Another assumption is that when otters are on land, usually asleep, their metabolic rate is the same as what we measured in their sleeping boxes in captivity, their RMR or resting metabolic rate in a thermo-neutral environment (average 3.2 W kg^{-1}).

With these provisions in mind, I constructed a model of the length of time per day which an otter of known or estimated weight would have to spend fishing in order to meet its daily energy costs, for different situations of prey availability, and for different water temperatures (Fig. 6.14). Just for information for physiologists, I assumed an average calorific content of fish of 4.5 kJ g^{-1} (Watt 1991), an assimilation coefficient for an otter eating fish of 0.7 (Costa 1982), and an RQ (respiration coefficient) of 20.1 (Bartholomew 1977; Williams 1986). Prey 'availability' can be expressed as weight of fish caught by an otter per length of time in the water.

In principle, the graph shows 'break-even points', the amount of time that an otter has to forage each day in order to make up its energy deficit in waters of different prey 'catchability' and at different water temperatures. The model shows that in an area and at a time when an

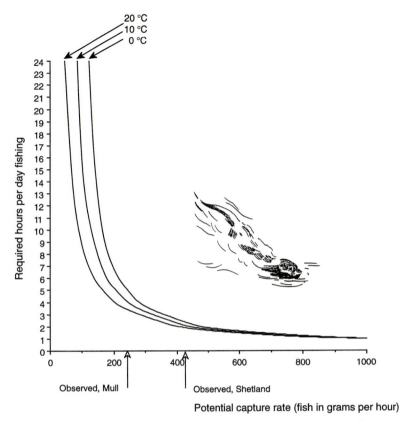

FIG. 6.14 Model of costs and benefits to otters of foraging in waters with different prey capture rates, at different water temperatures. 'Break-even curves', showing the length of time an otter needs to fish every day in order to make up for estimated energy lost. Arrows indicate observed fish capture rates in the sea on Mull (Scottish west coast; Watt 1991) and in Shetland.

otter can catch 600 g of fish per hour, it would have to fish for only about 1.5 hours to make up its estimated daily energy deficit. However, when otters can only obtain 150 g per hour they have to fish for almost half a day every day, at least in winter, when water temperatures are close to freezing. In summer, 5 hours fishing per day will suffice, because fishing in warmer water is that much 'cheaper'.

Despite the various assumptions I had to make for this model, it does show that the estimated quantities of fish needed to make up for loss of energy are reasonably close to the estimated quantities which otters actually take in the wild and in captivity, and this is encouraging. The model predicts that, largely dependent on water temperature, otters should take between 0.8 and 1.8 kg of fish per day for calorific contents of prey to match energy expenditure: this is, indeed, the range of food quantities we have estimated previously and which is mentioned in literature (p. 123).

The advantage of this energetic approach is firstly that it enables us to actually measure in the field what we think otters should be achieving. We can observe the time it takes an animal to catch a fish ('fish availability'), and we can estimate the size of the fish either visually and directly or deduce it from the size of fish vertebrae in the spraints of the animals. Water temperature can be measured directly, and with radio tracking we can now estimate the length of time per day which an otter spends fishing.

The graphs in Fig. 6.14 shows how vulnerable otters are when fish populations become very low, especially in colder regions; the animals can easily cope in an area where they can catch 200 g fish per hour: 6 hours' fishing per day in winter, and less than 4 in summer. However, if that prey availability were to decrease by half (which is not a very drastic reduction in fish populations), otter survival would become impossible in winter.

We can compare this with some actual observations on fishing success. For instance, in Shetland, where water temperatures range from 6 °C in winter to 12 °C in summer (Kruuk *et al.* 1987), there was a mean fish-capture rate of 460 gram per hour in summer (Nolet and Kruuk 1989), easily sustained at 2–3 hours' fishing per day. However, otters on Mull, along the Scottish west coast, caught only 290 g per hour in waters of 14 °C in summer (Watt 1991); there, a reduction of 50 per cent in fish availability could make otter existence impossible during the low temperatures in winter.

Fresh water gets much colder in winter than the sea, and Scottish streams may even reach sub-zero temperatures. In a set of preliminary (unpublished) observations of otters in the Dinnet lochs (Deeside, north-east Scotland) we saw the animals catch 83 eels over 28 h 50 min of observations, which is an average of 2.9 eels per hour in the water. This included the time spent in the water by cubs which were fishing and accompanied their mother. The mean estimated length of these eels was 32 cm, and from that a calculated mean weight was 80 g. Thus, the otters' hunting success was just under 232 g per hour, in water of between 1 and 14 °C.

At that rate of food intake, eating 'average' species of fish, an otter has to fish for 2.3 hours each day in winter temperatures to make up its daily energy deficit. Eels have a higher energy content than other fish (p. 135), therefore daily fishing periods will be reduced in areas where they are the main prey. But even there otters still have to work a long day to eat sufficient quantities, and the above calculations do not allow for other costs, such as reproduction, which are very high (Oftedal and Gittleman 1989). In rivers near our Dinnet Lochs study area, three otters with radio-transmitters were active for an average of 5.2 hours per day (5.6, 5.1 and 4.8 hours, total $n = 56$ days spread throughout the year) (Durbin 1993).

As I showed earlier (Chapter 4), it is just during the cold periods of winter and early spring that otters also take other prey such as birds and mammals. These relationships are being studied in detail at the time of writing, but already the preliminary observations suggest that otters are living rather close to the limits of their possible existence in fresh water.

THE DEVELOPMENT OF FORAGING BEHAVIOUR

In an excellent study on Mull, along the Scottish West coast, Jon Watt followed a number of young otters from their first fishing exploits in the sea into independence (Watt 1991, 1993). Cubs began to catch some of their own food when about 5 months old, slowly increasing this with age. When they were about 8 months old, they caught about half of their food themselves, and they were completely self-sufficient at about 13 months old. During this period, their foraging became increasingly independent of the mother; initially they would stay very close to her, often watching her under water whilst they floated on the surface with their faces submerged. When beginning to fish for themselves they would first dive synchronously with the mother, then at later age they would gradually dive further away from her. Under water, mother and cubs would often be very close as well and it was likely that cubs learned certain aspects of foraging this way, as do young foxes (Macdonald 1980b). There was also evidence of the otter mothers actually 'teaching' cubs to fish (p. 92).

With age, young otters also became more successful; overall in Watt's study on Mull the diving sucess of cubs was 13 per cent, of young independent (subadult) otters 24 per cent, and of adults 28 per cent. The same trend was apparent in the weight of prey captured per hour of fishing. Moreover, cubs were much more likely to take prey of inferior quality than adults; for instance, about 32 per cent of their 460 captures were crabs, whereas adults took those in only 3 per cent of 3052 captures, with subadults intermediate at 15 per cent (1255 prey). In general, foraging improved even after cubs were independent from their mothers, and the cubs continued to become more efficient until they were almost 2 years old.

One of the problems which young otters had was with their diving technique, or rather their breathing efficiency. The proportion of time they spent under water in every dive-interval, i.e. dive-to-recovery ratio for cubs was significantly lower than for adults, because the cubs' dives (for given depths) were shorter, although their recovery times were the same. If this is termed 'diving efficiency', then cubs became just as efficient as adults at the age of about 17 months, but amount of prey caught per dive continued to increase until they were 21 months old (Watt 1991, 1993).

Although we did not carry out a similarly detailed study on the development of fishing behaviour in the Shetland otters, the indications there were that animals matured considerably faster than in Watt's research on Mull. It was rare for otter cubs in Shetland to accompany their mothers until they were more than 10 or 11 months old, and with the odd exception, families split up well before the next breeding season (p. 87). Thus, most cubs in Shetland were born around June, and families split up before April. This was likely to be related to better feeding conditions (fish populations) in Shetland than along the coast of Mull, as estimated by fish trapping (Watt 1991).

Despite such differences between sites, it is justified to conclude that the development of foraging behaviour in individual otters takes a very long time, compared with other carnivores. Catching fish is intrinsically difficult for a terrestrial mammal, and the unusual amount of skill required is likely to be the reason for the long dependence of cubs on provisioning by the mother. This, in turn, will probably affect the breeding interval of the females, and as argued in Chapter 8, overall cub production.

FORAGING BY OTHER SPECIES

The 'normal' foraging behaviour, described here for the Eurasian otter, is common in all other species for which observations have been published, *mutatis mutandis*: it is typical otter behaviour. The sole exception to this is the sea otter *Enhydra*, and several other otter species have evolved additional behaviour patterns to catch prey. The characteristic otter foraging, in shallow water, for slow-moving, benthic, small fish, is in marked contrast with that of other piscivorous mammals such as seals and sealions, which go far and deep, and catch many mid-water as well as bottom-living fish, swallowing them under-water. But foraging techniques similar to those of otters are also found in mink (Dunstone 1993), and in the quite unrelated platypus (Kruuk 1993).

Foraging spotted-necked otters *Lutra maculicollis* in Lake Victoria, apart from applying the same diving-from-the-surface techniques as the

Eurasian species, also spent much time hunting fish in dense semi-aquatic vegetation, in daytime (Kruuk and Goudswaard 1990; see also Procter 1963; Rowe-Rowe 1977). Also in Africa the Cape clawless otter, *Aonyx capensis*, uses exactly the same diving techniques, but additionally, shows a characteristic difference from *Lutra* by probing under stones and vegetation with its long-fingered forepaws. In very shallow areas the head is held above water whilst feeling about with the paws, hunting for crabs (personal observation). In south-east Asia the small-clawed otter *A. cinerea* forages almost exactly like the Cape clawless, but there the smooth otter *L. perspicillata* fishes like the Eurasian species, with an additional technique: several individuals may co-operate and 'drive' fish (Kruuk *et al.*, 1994a).

The detailed descriptions of foraging in the North American river otter *L. canadensis* correspond to those of *L. lutra* (Melquist and Hornocker 1983; Melquist and Dronkert 1987), and for the marine South American species *L. felina* the same applies (Ostfeld *et al.* 1989). The giant otter *Pteronura brasiliensis* fishes the Amazonian rivers and streams, as described by Duplaix (1980) almost exactly like our otters in the north-east of Scotland: concentrating in small streams rather than large rivers, specializing in slow, bottom-living fishes.

But the sea otter *Enhydra lutra* forages differently from all the other ones, as it is different in almost every other aspect of its biology. It dives in waters much deeper than those used by *Lutra* species, commonly down to 40 m or more, frequently staying under water for more than 2 minutes (with a maximum recorded dive time of 246 s)(Riedman and Estes 1990). Prey, mostly molluscs, sea urchins, and crabs, are captured with the forepaws, and may be dislodged with the help of rocks: sea otters are well-known as one of the very few mammals which use tools (Hall and Schaller 1964; Kenyon 1969; Riedman and Estes 1990). Under water the captured prey is often stored under the loose flaps of skin in the axillae, and the animal may catch several before returning to the surface. The success rate per dive is over 70 per cent.

On the surface again a rock is often used as a tool to break open the shells, and a sea otter may carry a rock with it (again in its axilla), specially for this purpose. The variation in feeding techniques of this species is enormous, and individuals have learned to beg from tourists or fishing people, to open empty aluminium cans on the sea bottom in which they find octopus (McCleneghan and Ames 1976), to steal food from each other, and other methods. There appears to be much more individual variation in feeding habits in this species than we have recorded for the Eurasian otter, and there is evidence that pups learn individual foraging tactics from their mothers (Riedman and Estes 1990).

7

Thermo-insulation of otters in the sea and fresh water

Even when swimming in cold water body temperature is maintained around 38 °C, although otters experience cooling rates of about 2.3 °C h⁻¹. Body temperature increases prior to submersion, and animals leave the water at approximately overall mean body temperature. Differences in water temperature had no discernible effect on body temperature, or on time spent swimming. For thermo-insulation otters rely almost entirely on their fur and air trapped inside it, and fur maintenance (grooming, rolling) is described and compared with that of some other species. The most elaborate fur maintenance is found in the sea otter. When Lutra *is living in the sea it is dependent on pools of fresh water for fur maintenance; often these pools are found deep inside the holts. Thus, the distribution of otters along sea coasts is dependent upon the distribution of fresh water.*

INTRODUCTION

The first, single most important ecological characteristic of otters, the one that everybody knows about, is the fact that they live in cold water. This creates big problems, and the present chapter discusses the extent of these problems, the adaptations in otters to overcome them, and the effects of this on their survival and distribution.

The normal body temperature of mammals is around 37–39 °C, and otters are no different from others in this respect: theirs is usually around 38 °C in our own observations (see below). The big challenge to the animal is to maintain temperature in the water, and it spends a great deal of energy in doing this. Another semi-aquatic species of mammal, which differs substantially from the usual mammalian pattern, is the platypus, with a body temperature of only about 32 °C (Grant and Dawson 1978; Grant 1983). This should lighten its thermo-regulatory burden somewhat (Kruuk 1993), but otters have not taken to this particular solution, and they have to maintain their high body temperature under very testing conditions.

In Britain water temperatures are usually well below 20 °C, in winter often close to freezing point in fresh water, and between 6 and 8 °C in the sea (Fig. 7.1). Our studies on otters in lochs in Scotland, where the observations on otter body temperature reported below were made, measure winter water temperatures down to 0.7 °C, in summer going up to about 23.5 °C, with an annual mean of 9.4 °C (Kruuk *et al.*, in preparation); in streams the temperature goes down to –0.7 °C in winter. One may expect similar conditions or even colder ones in the northern areas of the otter's range in continental Europe and Asia. Even in the

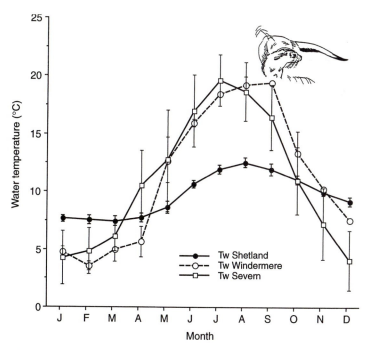

FIG. 7.1 Water temperatures in Britain at different times of the year, ± standard deviations. Shetland 1982–5; Lake Windermere 1958–66; River Severn near Tewkesbury 1980–5. (After Kruuk *et al.* 1987.)

southern parts of its range the species will frequently meet water temperatures not too much above freezing.

In itself this would not be so much of a problem if the conductivity of water were not so high. It is 23 times the conductivity of air (Schmidt-Nielsen 1983), and an otter swimming in 'average' water will lose heat very fast, unless something is done about it.

In this chapter I will discuss changes in body temperature, the importance of thermo-insulation, and aspects of otter behaviour and other mechanisms whereby insulation is maintained. I will describe experiments with captive otters to demonstrate the different effects of sea water and fresh water in all this. We can now demonstrate how these needs for thermo-insulation are related to the animal's ecology and distribution, and this will be compared with what happens in other species of mammals, such as the sea otter *Enhydra lutra*.

BODY TEMPERATURE (T_b)

To let body temperature fall below a given value may cause permanent damage to the system, and must therefore be prevented. Insulation helps, and this will be discussed below, but this can never be completely effective, and physiological and/or behavioural adaptations are also required.

When an otter is exposed to cold water it could follow several possible strategies or combinations: it could

1. increase T_b before entering, thus forestalling a damaging decrease, or

2. maintain a fairly steady T_b throughout, with increases associated with increased activity, or

3. tolerate a small decrease in T_b and leave the water when T_b becomes critical, or

4. increase the intensity of activity whilst in the water, thus generating a higher T_b.

In addition, there are ways in which otters could possibly modify their behaviour to allow for the effects of lower water temperature (T_w): for instance, they could spend less time in the water. However, if this is to be done without affecting the time spent actually searching for prey underwater, the only way to achieve this would be to vary time on the surface between dives, the dive recovery time. If this were happening, we should see shorter surface times (and dive intervals) in colder waters. We tested all these predictions on body temperature, water temperature and related diving behaviour in a project in conjunction with the study on oxygen metabolism described in Chapter 6.

When radio-tracking otters in freshwater lochs in Scotland, we decided, for four animals, to use transmitters which gave information on body temperature, so we could relate this to what they were doing, and to the temperature of the water in which they swam. Such transmitters are inserted intraperitoneally, like the ordinary ones described on p. 19 (so we could obtain core body temperature), and they are about the same size, but their working life is about half (4 to 6 months). The interval between radio pulses is a measure of body temperature, and this can be recorded and analysed on special equipment.

Overall, in 1770 recordings on the four otters (on land and in water), the mean body temperature T_b was 38.10 °C, but there were small variations, for instances between individuals and for different activities. The lowest recorded T_b was 35.86 °C, the highest 40.37 °C, a difference of 4.51 °C. The mean of individual body temperatures for inactive otters was 38.14 °C, in one individual as high as 38.71 °C, in another as low as 37.56 °C. There were no discernible temporal patterns in daily fluctuations of T_b.

In 59 observations of otters in the lochs, whilst we were recording body temperature from their radio signals, we collected data on T_b when they entered and left the water, the length of time they had stayed in, and we also knew the water temperature (T_w). From this we could calculate the total loss of T_b in the water, and also the rate of loss, then relate all these variables to each other by using multiple regressions. The main conclusions from all this were, firstly, that there was a rate of loss of T_b in the water of on average 2.3 °C h^{-1}. There was a mean total decrease of T_b per foraging bout of 1.1 °C. These data were similar to the results of MacArthur and Dyck (1990) for young beavers.

Secondly, T_b was high when an otter entered the water, significantly higher than average T_b (35 observations, $p < 0.001$, sign test), as is known also for other species such as the muskrat Ondatra zibethicus (MacArthur 1979). Probably, this is caused by increased activity on land before diving in. Whilst swimming, T_b decreased, and at the end of the swimming bout it was not significantly lower than average T_b ($p > 0.1$). Fig. 7.2 shows a typical observation of the course of body temperature during a foraging trip.

Thirdly, we found no significant effects of water temperature T_w on any of the variables related to body temperature, such as T_b at the start or at the end, the difference between the two, or the rate of loss in T_b. This was despite the fact that T_w varied during our observations between 2

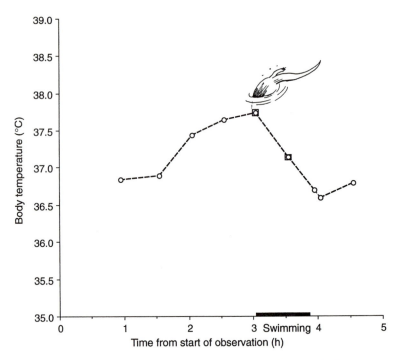

FIG. 7.2 Core body temperature of a wild otter in its natural habitat, before, during and after aquatic activity (example). ○ = otter on land, □ = active in water; bar indicates the length of period active in water. Radio-telemetry data, water temperature 6.7 °C. Dinnet Lochs, north-east Scotland.

and 16 °C. Also the length of the swimming bout was unaffected by T_w, unlike beavers or muskrats, which swim less when the water is cold (MacArthur 1979, 1984; MacArthur and Dyck 1990). However, it was possible that otters were more vigorously active when the water was cold; we could not quantify that. In fact, we should not expect water temperature to affect the length of the foraging period, because that would have an immediate effect on foraging success in otters: this is different in beavers and muskrats, which forage mostly on land.

However, we did find a small but significant effect of T_w on the dive interval, though quite the opposite of what we expected: in summer dives were shorter, as were surface times and dive intervals. I think, though, that the explanation for this is more likely to lie in the environment changing with the season: for example water levels were somewhat lower (dive times are correlated with depth of water; see p. 173).

To return to the hypotheses on otters' body temperature when swimming, the observations suggested clearly that although there was a substantial drop in T_b during aquatic activity (at the rate of 2.3 °C h^{-1}), otters anticipated this by entering the water with a T_b which was higher than average (probably after increased activity on land), and they came out with a T_b which was approximately average. It was not possible to say, however, whether animals could have stayed in longer, using increased activity in water to boost their T_b, and still come out with 'average' T_b. Differences in water temperature had no effect on any of this.

Thus, core body temperature is maintained by otters within limits which are more or less normal mammalian values. A high metabolic price is to be paid because of the conductive properties of the cold, aquatic environment (Chapter 6), if the animal is not to make major concessions to foraging strategies.

THERMO-INSULATION IN OTTERS

Seals, which may live in the same places as otters, are protected against cooling by a subcutaneous layer of blubber some 7 to 10 cm thick (Irving and Hart 1957). Otters, however, have no blubber at all, and very little fat: to them, the fur is all-important, and blubber would have some serious disadvantages (see p. 195). In the 'fattest' otter dissected, it was found that about 3 per cent of its body weight consisted of fat as adipose tissue (Pond 1985), and that was mostly mesenteric, that is inside the body cavity. This compares with a seasonal presence of over 30 per cent of body weight in thick layers of fat in another mustelid, the badger *Meles meles* (Kruuk 1989a). There is some fat under the skin of otters in the loins, and especially around the base of the tail, but even there small muscles are to be found in the layer between the fat and the skin, and in general fat cannot contribute much to thermo-insulation.

It has been shown that fur is far less efficient as an insulator than blubber (Tarasoff 1974), yet otters are entirely dependent on it for protection against cooling in water, and their coat is adapted specially for this purpose. Under an outer layer of guard hairs, each up to 4 cm long, there is a layer of under-fur, an extremely dense mat about 1 cm thick, through which the skin is quite invisible even when one tries to penetrate the fur. When examining an otter skin closely, the under-fur gives the impression almost of being the skin itself, so dense is the hair. One of my students, Addy de Jongh, looked at this through an electronmicroscope, and counted the hairs in bundles of about 20 or 22 under-hairs around each guard hair. There were 20 to 30 such bundles per mm^2, or about 50 000 hairs per cm^2. However, sea otters have even thicker fur, at more than twice this density (Kenyon 1969; Tarasoff 1974).

During a dive, air is trapped in the under-fur, and this is important for thermo-insulation. For instance, it has been shown in polar bears (*Ursus maritimus*) that the thermal conductivity of fur is 20 to 50 times greater when wet than when dry (Scholander *et al*. 1950). It is therefore vital for otters to maintain the air-holding capacity of fur, even at the cost of considerable effort; it has been demonstrated for the Pacific sea otters that the air-holding capacity of fur is lost very easily after even moderate fouling of the pelt, or by contamination with oil (Costa and Kooyman 1982).

Our otters spent quite a lot of time grooming themselves (Fig. 7.3), and especially rolling on grass and seaweed. Bart Nolet, a Dutch student working in our Shetland team, found that three Shetland otters with radio-transmitters spent 6 per cent of their time grooming, and the length of each grooming bout was correlated with their previous behaviour in the water (Nolet and Kruuk 1989). The correlation with length of time fishing was not significant (a correlation factor $r = 0.18$), but there was a significant correlation when the depth at which the animal had been fishing was taken into account, by comparing length of grooming bout with time in the water multiplied by the depth of dives ($r = 0.25$, $p < 0.05$). This suggested that diving, especially deep diving, stimulated grooming. In later years I have become convinced that the radio-transmitters, which in Shetland we attached to the otters with a small harness or collar, affected these results; otters without them spent less time grooming and rolling. After the experiments with captive otters, which I will describe below, we also know why these radio

FIG. 7.3 Grooming (male, in captivity).

attachments should have affected the animals: they stopped otters grooming some parts of their bodies. Anything which interferes with the maintenance of the natural functions of the fur affects the otters' behaviour. Nevertheless, the effect of previous diving on grooming was likely to be real, and not something solely caused by the attachment of the transmitter.

Whether otters living in fresh water spent just as much time on fur maintenance as those in the sea we do not yet know, as they are more difficult to observe continuously. I would predict that freshwater otters groom and roll much less, because of the effect of salt water on the pelt. This effect of sea water, and the otters' reactions to it, opened our eyes to many of the problems of thermo-insulation. As early as 1938 Richard Elmhirst, who was at the time Director of the Marine Biological Station in Millport, near Glasgow, described in detail his observations on otters' use of small freshwater pools along the coast for rinsing off the sea water. Having seen this behaviour in Shetland, I realized that these basic observations explain much of the otter's distribution along coasts and their behaviour.

The effect of salt water on otter fur was analysed by another student, David Balharry, who did experiments in a project with our captive otters at the Institute of Terrestrial Ecology in Banchory (Kruuk and Balharry 1990). Two adult females, Tubby and Yoyo, were kept in an enclosure of about 5 by 14 m, with a large swimming pool. There was a small puddle under a tap where they could always drink if the pool was dry. For the experiment in which we tried to simulate conditions along the sea coast, we gave the animals an additional large pool in the enclosure, a circular one made of fibreglass, which I shall call the feeding pool. We sometimes filled the feeding pool with sea water, collected directly from the North Sea, and sometimes with fresh water from the tap. Thus, we could have the otters using the feeding pool with sea water, whilst they did or did not have access to the freshwater swimming pool, or we could make the otters use the feeding pool full of fresh water, again with or without access to the freshwater swimming pool. Most of the experiments were done in October and November, so the water was fairly cold, between 1 and 6 °C.

Tubby and Yoyo were fed five times per day, each feeding period lasting 25 min. During one feeding period, each animal would get five pieces of haddock, which were thrown at 5-minute intervals into the feeding pool. We measured several different aspects of the otters' behaviour during and in between these feeding times, and there were striking differences in what the otters did when they were being fed in sea water or fresh water.

First, during feeding in fresh water they spent somewhat more time in the water (on average 263 s, measured over 20 periods, compared with 217 s for sea water, in 40 periods), but this was not statistically significant. Especially striking was the observation that the otters paid many more visits to the alternative, freshwater swimming pool when they were being fed in sea water. This happened during 35 out of 40 feeding periods in salt water, but only once during the 20 feedings when food was given to them in fresh water. The difference was highly significant ($p < 0.001$, Mann-Whitney U-test).

The behaviour of both Tubby and Yoyo changed quite dramatically if they were being fed in sea water without having access to the swimming pool for a refreshing dip. For instance, in between feeds during any one feeding period, the animals used to dive in and out of the feeding pool, perhaps to check up if any food were left, or just out of excitement. They were much more reluctant to do this if they had not had their freshwater dip for one or more days (Fig. 7.4). And if they had been fed in sea water for four days or longer, without having access to fresh water for swimming, the animals would be clearly miserable, sitting at the edge of the feeding pool, shivering and reluctant to enter the sea water even during actual feeding, and even when the air temperatures were not particularly low. Of course we then had to stop the experiment and turn on the tap for the freshwater swimming pool, which was the signal for Tubby and Yoyo to dive in, playing and splashing.

Often an otter gives itself a good shake after it leaves the water (Fig. 7.5). In our experiments we saw that this, too, was very much affected by sea water: in the absence of fresh water for washing, Tubby and Yoyo shook themselves significantly more often, and clearly it was the sea water which caused the increase (Fig. 7.4). The same pattern emerged for grooming and licking their fur, which the otters did about twice as much when they had sea water in their feeding pool, compared with fresh water.

These are just some basic quantitative effects of sea water on otters, but to anyone simply watching the two animals when there was only sea water in their enclosure, it would have been obvious that something was amiss even without these statistics. An otter emerging from sea water, after several days without a good rinse in fresh water, appeared to be quite soaked through, with the pelt hanging heavily around it—totally different from the smooth but fluffy coat we had come to accept as normal. In fact, we also quantified this phenomenon by making the animals enter and leave the feeding pool over weighing scales, so that we could weigh the quantity of water clinging to their fur. This was somewhat rough-and-ready, but the scales did show that (a) the longer

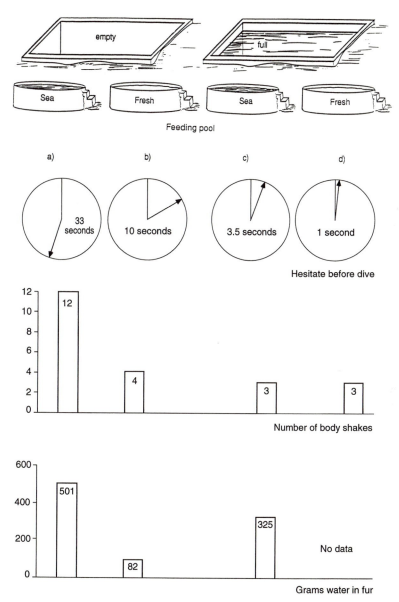

FIG. 7.4 Results of observations on two otters in captivity, in which they were fed in sea water (a and c) or fresh water (b and d), and also had access to fresh water (c and d) or not (a and b). The otters hesitated longer to enter sea water when no alternative fresh water was present (40 observations; Mann-Whitney test $U = 274.5$, $p < 0.001$), they also shook themselves more (40 observations; $U = 156.5$, $p < 0.001$), and ended up with more water clinging to their pelt (16 observations; $U = 3$, $p < 0.001$). (Data from Kruuk and Balharry 1990.)

FIG. 7.5 Yearling female otter shaking water out of the fur during coastal foraging.

the animals stayed in the feeding pool, the more water they absorbed in their pelt and (b) during seawater feeding the amount of water absorbed by the pelt was significantly greater (Fig. 7.4).

This last result suggests that sea water interferes with the capacity of the fur to hold air under water, and it soaks up water instead, seriously jeopardizing the insulating function of the pelt. Apparently what happens is that when sea water dries in the pelt, salt crystals are formed along the length of both guard hairs and under-fur, with many hairs sticking together in small bundles. The salt could interfere with the lipid secretions of the skin glands on the hairs, which are instrumental in retaining the air layer (Tarasoff 1974). Possibly, the crystals could also interfere with the surface structure of the individual hairs, the scales.

To confirm these results in the laboratory, some experiments were done with pieces of otter skin collected from road casualties, measuring the temperature at several points inside the fur with small thermocouples. The pelt was stretched over a copper plate at 35 °C (just below body temperature) with cold (6 °C) water flowing over the outside. The temperature near the skin inside the fur was 29.1 ± 1.1 °C ($n = 10$ observations on four pelts), which showed that heat permeated the skin from the inside, but not much got lost through the fur. After the pelt had been rinsed and dried twice in sea water, there was no significant change in this, but after five times washing in sea water, the temperature in the same sites inside the fur dropped to $23.0 \pm 1.1\,°\mathrm{C}$ ($n = 15$), a significant decrease ($p < 0.01$, Mann-Whitney test). There was no significant decrease in a set of control pieces of otter skin, which had been washed in fresh water. This demonstrated the thermo-insulating effect of the fur, and the substantial havoc played by sea water.

Otters do not have any obvious morphological adaptations to facilitate cooling, such as large, vascularized appendices (ears, flippers), or behaviour patterns such as panting. Clearly, keeping heat in is more of a problem than disposing of it.

THERMO-INSULATION IN OTHER SPECIES

Several species of mammal have to cope with particular problems of thermal insulation in sea water. Seals and whales solve these with a layer of blubber, but that is a luxury which only the larger mammals can afford. Blubber is an extremely efficient insulator, much more so than fur (Tarasoff 1974). However, it also has disadvantages, especially its low specific weight: just as if one were to attach a large cork to an otter, blubber would mean harder work for an otter to dive with. Seals do not appear to have this problem, because they dive with very little air in their lungs (unlike otters), and their overall specific weight is therefore greater than that of water, even with the protective envelope of blubber. A seal can submerge in its vertical sleeping position in the water with its head up, just by expiring; it slowly sinks, without even a ripple showing on the water surface. An otter has to make a big effort to go down, diving head first, tail up, with a lot of propulsion from the back legs.

A disadvantage of fur as an insulator is that it relies on a captured air layer for its effectiveness, and this compresses with depth. On the other hand, any animal which is equipped with a thick layer of blubber will find itself seriously encumbered when moving about on land, because of the increase in weight and inflexibility (Estes 1989), and for this reason alone it would not be suitable for semi-aquatics such as otters, mink, water shrews (*Neomys fodiens*), and others.

For all the smaller species, of the same size as otters or smaller, insulation with fur is the only solution to a problem which is even more serious for them than for large mammals. Their troubles are exacerbated not only because small bodies have a smaller thermal capacity (that is they can hold less heat), but also because the relative surface area is greater in a small animal. If, given two mammals of the same relative proportions, one has a weight which is one-tenth that of the other, its surface area will be about one-fifth, relatively twice as great.

In other words, a small animal in cold water will cool much more rapidly than a large one with the same thermo-insulation.

One of the most interesting behavioural adaptations to cope with thermal insulation, and with the problems of fur in sea water, is found in the sea otter *Enhydra* in the northern Pacific. It is the most aquatic of the carnivores, spending almost all of its life in sea water, sleeping, feeding, grooming, from birth to death (Kenyon 1969). Many sea otters live in arctic temperatures; nevertheless the species has no blubber, but relies entirely on the thermo-insulation provided by the fur. How do these animals manage to keep their fur in trim, obviously without going through all the rigmarole of washing in fresh water?

This is achieved by several means. First, the fur of sea otters is much denser than that of Eurasian otters (Tarasoff 1974), and secondly sea otters are larger (two to four times the weight of *Lutra lutra*) so cold water affects them relatively less. Thirdly, in order to cope with the heating demands, their metabolic rate is two or three times higher than that of similar-sized mammals (Costa and Kooyman 1982; Morrison *et al.* 1974), and this means they have to eat more. Fourthly, as sea otters leave the water far less frequently than Eurasian otters, the sea water does not dry out in the pelt and there will be less salt encrustation in the fur. Nevertheless, fifthly, sea otters have to keep their fur scrupulously clean by grooming a great deal every day; if the hair gets only slightly oiled, or if the animals get dirty under conditions of captivity, their thermo-insulation is breached, and they develop pneumonia very quickly. This has been shown, initially more or less accidentally when animals died in captivity (Kenyon 1969), and later by other scientists in deliberate experiments which studied the effects of oil pollution at sea on sea otters (Costa and Kooyman 1982).

Finally, as a sixth adaptation to their thermal problems, what I found especially exciting was the finishing touch of the grooming behaviour of sea otters. *Enhydra* spends 1 to 2 hours each day grooming, mostly or entirely when floating in the water. They squeeze and rub their fur with their forelegs, licking it all the while, and as they have a particularly loose skin they can reach virtually all parts of the body in this way. At the end of an aquatic grooming session the floating sea otter shows a characteristic behaviour pattern, blowing and whipping the air back into the fur (Kenyon 1969). In contrast, *Lutra* never grooms in the water, but does this on land rubbing on grass and seaweeds, drying the fur in the air and relying on this and the use of fresh water to rid the pelt of salt, and to allow air back into it.

In sea otters most heat flux is conducted through the large rear flippers, which are sparsely furred and heavily vascularized (Iverson and

Krog 1973; Morrison *et al.* 1974; Costa and Kooyman 1982; Estes 1989). They may be used not only for cooling, but also as 'solar panels' to absorb heat, with the animals floating on their back, flippers out of the water (Tarasoff 1974).

There are other species of otter which live in the sea as well as in fresh water, for example the North American river otter *L. canadensis* which also occurs along the coasts of Alaska (Melquist and Dronkert 1987). A large African species of otter, the Cape clawless or *Aonyx capensis*, lives mostly inland but also occurs along the ocean coasts of South Africa where it is associated with small freshwater pools and streams to bathe in (Arden-Clarke 1986; Verwoerd 1987). To me it was fascinating to see this species, of a different genus, on another continent, eating different prey, but using the marine habitat with exactly the same constraints of freshwater pools as the Eurasian otter. The much smaller African spotted-necked otter *L. maculicollis* is known from freshwater habitats only. In Chile the marine otter *L. felina* is found, apparently only along the ocean coasts and in areas reportedly devoid of any fresh water (Ostfeld *et al.* 1989). It should be interesting to study how that species copes with the problem of fur maintenance.

Several more small mammals apart from otters have been able to colonize freshwater habitats. In Europe these include amongst the carnivores species such as the mink, amongst the insectivores, the water shrew and the Pyrenean desman (*Galemys pyrenaicus*), and amongst the rodents species such as the beaver (*Castor fiber*), muskrat (*Ondatra zibethicus*), coypu (*Myocastor coypus*), water vole (*Arvicola terrestris*), and others. They all use a layer of air in the fur as thermal insulation. However, none has been able to adapt to a marine environment, with the exception of the mink which was introduced in Europe from North America: it is common along the coasts of Norway and in some areas of the Scottish west coast (Dunstone 1993). In Norway I noticed from tracks in the snow that coastal mink also washed in fresh water, often in the same sites as otters. As an example from the past, along the east coast of North America the sea mink (*Mustela macrodonta*) lived exclusively in marine habitats, until it was exterminated during the last century. It was much larger than the common mink, larger also than the otter (Dunstone 1993).

However, it seems that the extra problems of thermo-insulation posed by sea water in the fur of small mammals have been impossible to overcome by most species. The otter is one of the few exceptions, and it has been very successful in colonizing some coasts. Nevertheless, it still remains highly dependent upon the vagaries of the distribution of fresh water along the sea's rocky shores.

WASHING IN POOLS AND HOLTS ALONG SEA COASTS

Having demonstrated the importance of freshwater baths to otters in captivity, it was important to know exactly where the animals satisfied this need in the wild. It is not often that one actually sees otters wash and swim in fresh water along coasts as we did in Shetland, although it happens frequently. We knew this because of the well-beaten otter paths to the washing pools, and from following tracks in the snow. I would guess that an otter living in the sea washes in fresh water at least once every day, therefore one still has to be lucky to see it (especially since it often happens underground, or in darkness). The following were typical observations.

April 1986, in Lunna, Shetland. Mid morning, and I am following one of our ear-tagged males, Yellow-lug. He is a big, territorial and resident animal swimming along the rocky coast, occasionally diving, but mostly along the surface. The otter is surprisingly conspicuous, with his tail floating behind him, swimming slowly about 10 metres out from the rocks. He makes for a protruding rock, lands, spraints—then, instead of diving back into the waves again, hops over a few large boulders, and rapidly climbs the steep grass and gravel bank. The animal spraints again, then follows an otter path, in his typical clumsy gait, for just about forty metres, stops, spraints on a large heap of older otter spraints and slips vertically down a narrow hole in the peat. Although I am about 100 metres away, I do not need to be closer because I know the place and I have my binoculars: the hole is full of water, and it goes down vertically for about one metre. Another hole in the peat, also full of water, is connected with the first one underground about five metres further, and Yellow-lug pops up from that about twelve seconds later. He shakes vigorously, with a lovely halo of spray around him, then bobs back again to the sea, down the sandy cliff, over the rocks and into the surf, continuing his journey in the same direction.

An early Shetland morning in February 1987. A young female, Weibka, swims along the coast of West Lunna Voe, accompanied by her two cubs, now about half a year old. The party lands on a rocky shore at the bottom of a grassy slope, and the otters run a few metres higher up to a flat piece of ground, settling down to roll and groom, initially ignoring the big spraint site about two metres from them. After 17 minutes Weibka gets up, walks to the spraint site, spraints and returns again to her rolling lawn. The whole party continue to roll, groom, sometimes sleep for a bit. Another six minutes and they get up, walk down to the rocks on the shore and all three drink, from a tiny puddle between the stones, fed by a trickle of water running down from the peat above. The puddle is hardly the width of an otter's head, just one of those ordinary seepages of which there are thousands along the shore here. Weibka dives again, a ripple and a splash, then is followed by her offspring. Three dives on and the mother lands a big rockling, about 30 centimetres long; all three eat from it, although most goes to the cubs. One of the youngsters walks up to the spraint site and spraints; then all make their way up the grass bank, past the previous rolling site – and suddenly they are all gone from view, down a narrow hole full of peaty water.

The otters pop up again from a hole at the other side of a small peat bank, two metres away, galloping up the grass slope, soaking wet as they are. I see them for only a few more seconds, before they disappear down the entrance of a holt, in another peat bank. When they come out again, more than half an hour later, they are as perfectly dry as if water had never touched them. They gallop again, down and down, straight into the water where Weibka dives, in her seemingly endless quest of provisioning the two hungry mouths, with the cubs just behind her.

The significant part of this last observation, the watery dip in the peat hole, could so easily have escaped attention, as I am sure happens often.

There were many potential otter bath tubs along the shores of our study area at Lunna, some no more than a small, deep puddle between the rocks (Fig. 7.6), or a pool in the tiny burns which come trickling

FIG. 7.6 Freshwater washing pool (near the Scottish west coast, Arisaig). Note the spraint sites.

down the peaty slopes, others which were proper small lochs. The favourite ones appeared to be either the rocky pools close to the shore, or the otter-sized holes in the peat (Fig. 7.7), which went vertically down into the groundwater and were obviously made by otters themselves at some stage in the distant past. Some of these latter tunnels were neatly U-shaped, so the animals could go in one end and out the other. Or the otters turned upwards again after going down for 0.5 m or so, into a dead end underground. All these varieties were clearly recognizable as otter bath tubs; there was rarely any live vegetation in them, and there were often otter tracks, worn vegetation near the edge, rolling sites, and, most telling of all, spraints.

This abundance of spraints near freshwater pools and burns along the coast in itself testified to the importance of these places to otters. Sometimes there were complete hummocks consisting entirely of otter spraint, and especially on the Scottish west coast these were often 15 cm high or more. Elmhirst (1938) called them 'otter seats', very conspicuous and announcing that the site was a focal point for otters. As argued in Chapter 3, spraints function to communicate to other otters that a resource is being utilized; the effect should be that other otters avoid it. In this case, the sprainter keeps its bath water to itself, avoiding 'pollution' by others, and other otters would do well to take their cue from the spraints and bathe elsewhere, in unadulterated water.

There were many other otter baths along the coast which were not easily seen in the field. These were the ones located inside the holts,

FIG. 7.7 Freshwater washing site in peat: a 'squeeze hole' in Shetland. The tunnel extends for about 0.8 m under water.

sometimes visible in the entrance (Fig. 7.8) but often deep down; it needed a major excavation to expose them (Moorhouse 1988). This often explained why some of these otter holts were located in particular, curious positions: for instance on top of hills (Fig. 2.11, p. 45), or high up in the hills at the bottom of some small valley, perhaps a kilometre from the sea. Invariably, there was a dip in the rocks or in the impermeable clay underlying the peat, in which water could collect. The extensive tunnel system of a holt made in the peat above would have one or more branches which widened into a bath tub, with smooth well-worn sides and rather muddy water in it. Many holts had running water in them, when they were situated exactly on one of the small drainage lines carrying water from the peaty slopes down to the sea, or when the otters lived in underground flow systems which they had enlarged into a holt (Moorhouse 1988; see also p. 46). Frequently there was nothing to tell from the outside that such washing facilities were present in the holt— just simple entrances somewhat larger than those of a rabbit warren with no water anywhere nearby.

During dry times of the year, otters abandoned the holts in which the water dried up; of seven holts excavated, the three dry ones were not in use at the time, but the four with full bath tubs were well-frequented (Moorhouse 1988). An important feature in the underground bath tubs,

FIG. 7.8 An otter holt along the Shetland coast, with fresh water outside (as well as inside). Note the spraint sites.

as well as in the more exposed washing places, was that there were often one or more sites which allowed otters to wring water and air from their fur, usually by squeezing themselves through narrow gaps or underwater tunnels; we used to call the more obvious ones 'squeeze holes'. One also finds these squeeze holes in freshwater habitats of otters, probably contributing to fur maintenance in a similar manner. But they are nowhere near as common near freshwater lochs or streams as they are along sea coasts, even where there are probably just as many otters.

Thus the ostensibly simple requirement of otters, to be able to maintain their body temperature when swimming and foraging, appears to have major consequences for their behaviour, for their choice of habitats, and ultimately for their distribution and therefore numbers.

8

Populations, survival, and mortality

Different ways of assessing numbers of otters are discussed, including direct observation, spraint surveys (which should be treated with caution), stratified density sampling of otter holts, and the use of radionuclide-labelled spraints. Densities were found of one adult per 1–2 km of coast (Shetland) or 15 km of stream (north-east Scotland). Mortality patterns and age structure of populations are discussed; otters die relatively young, with a mean adult life expectancy of 3.1 years, and mortality rate gradually increases with age. Mercury is the only present-day pollutant on which we have information, which is likely to play a significant role in causing mortality amongst Scottish and Shetland otters. Prey availablity appears to affect both mortality and recruitment very strongly, but in different ways: mortality in spring, recruitment in summer. Females may deliberately abandon young cubs.

INTRODUCTION

Whether one is interested in behavioural ecology, the evolution of behaviour patterns, or the conservation of a species, a knowledge of population dynamics is vital. This includes especially an understanding of numbers in different areas, the mechanisms of population regulation, the relative importance of various causes of mortality, and the factors which limit reproduction and recruitment. Such information is scarce for otters, and in this chapter I will discuss the data now available, and draw tentative conclusions on some of the causes of mortality and success of reproduction. Clearly there are gaps in our knowledge, and I will explore some of these.

Perhaps the main reason for the lack of information on otter populations is the difficulty of assessing population size in a highly secretive,
often nocturnal, and sparsely distributed species. However, recently we
have made some progress with this, first in Shetland by using an indirect
sampling method (Kruuk et al. 1989), and later in freshwater areas by
following methods first developed for badgers (Kruuk et al. 1980, 1993).

One of the few advantages which are conferred on studies of mammal
populations over those on birds is that mammals can be aged accurately,
at least after death, by counting incremental rings in the dentine of
teeth, a method developed for otters by the Norwegian scientist Thrine
Heggberget (1984). This has enabled us to study population composition, patterns of mortality, cumulative effects of pollutants, and other
important effects and differences.

NUMBERS OF OTTERS

Methods, general

The numbers and densities of many species of mammals and birds can
be estimated either directly, that is visually, or from the number of territories assessed by observing territorial displays (in birds, for example
Davies (1992)), or from the pattern of distribution, as in foxes (Harris
and Smith 1987), or badgers (Cheeseman et al. 1987; Kruuk and Parish
1987). Otters present difficulties. In Shetland, where they are diurnal, it
is possible to recognize almost all individuals over stretches of coast of a
few kilometres, after intensive observation over a few weeks there.
However, those kinds of field conditions are exceptional, and in general
it is impossible to get any idea of otter numbers without using some indirect method of census.

The most common census method, on which most of our assumed
knowledge of otter numbers is based, and from which we deduce most of
the pattern of their decline and fall in various countries, is the survey of
otter spraints (Mason and Macdonald 1986). Spraints are often the only
evidence of the animals' presence, they can be quite conspicuous, and
frequently they are found in convenient spots such as under bridges.
Surveys are based on presence or absence of spraints, and one of the
standard methods uses a 600 m stretch of bank: if no spraints are found,
one scores a negative, and so on. At the end of a survey the percentage of
'positive' sites is taken as a measure of the strength of the otter population. In earlier days such surveys were used to estimate actual numbers
of otters; for instance, it was concluded from spraint distribution that in
1975 there were 36 otters in Suffolk (Mason and Macdonald 1986).

However, there are serious problems with such deductions. Clearly, when spraints are present there are otters, and in general, in areas where there are many spraints there are likely to be more otters than in regions where spraints are few. But otters frequently defaecate when swimming, as I have seen many times both in the wild and in my captive animals, and the spraints we find on the bank have a specific biological function apart from elimination (Chapter 3). Thus, absence of spraints does not necessarily mean absence of otters. In our study in the north-east of Scotland where we follow otters with radio-tracking, we find that there are several areas where the animals spend a great deal of time (in reed marshes for instance, far away from open water), but where despite intensive effort we can never find any spraints at all. We tried hard, because we wanted to study the otters' food in those sites.

Not only does absence of spraints tell us little about otter activity, but also when spraints are present numbers do not always correlate with otters (Kruuk *et al*. 1986). In Shetland we surveyed 21 350 m stretches of coast for spraints, whilst we also watched for otter visits, and no significant relationship was found. This was a simultaneous comparison of different areas, and in some of these coasts we found many spraints, elsewhere very few, independently of the numbers of visits by otters and time spent by the animals. In an apparent contradiction later, when watching actual sprainting behaviour, I saw otters deposit more spraints in sections where they also spent more time. The most likely explanation for the discrepancy was that many spraints were deposited on seaweed below the high-tide mark, and these were soon washed off again. Such problems would not arise in freshwater areas, along rivers.

On top of these inconsistencies comes the problem of seasonality: otters spraint on land up to seven times more in winter than in summer (Mason and Macdonald 1986; Conroy and French 1987; Kruuk 1992), and this annual fluctuation is by no means regular, with peaks, troughs, and main increases or decreases occurring during different months (Conroy and French 1987). This makes sprint density, even at a given time of year, a tenuous measure for utilization of an area by otters.

These problems have been argued at length in two papers in *Biological Conservation* (Mason and Macdonald 1987 *versus* Kruuk and Conroy 1987). I can only suggest that one should accept statements on otter density with great caution if they are based on spraints, but it would be unwise to disregard evidence such as that from repeated national sprint surveys. In these surveys experienced field workers, using a somewhat vague correction for seasonality, find repeatedly that in recent years spraints occur in an area which is gradually enlarging, in Wales and from Wales further into England (Crawford *et al*. 1979;

Andrews and Crawford 1986; Strachan *et al.* 1990), and there is the clear suggestion that otters are making a come-back after being absent for decades. How strong this come-back is, is still open to surmise.

All these arguments led us to decide that spraint abundance would be too complicated and unreliable an index for otter numbers and distribution, both in Shetland and elsewhere. We resorted to two alternative procedures to arrive at a figure for otter density, one to be used in Shetland and one in freshwater areas. The main reason for two distinct methods was the difference in life style of the animals: in Shetland the otters use holts on a daily basis, whereas along rivers and lochs they usually sleep in the open, even in winter. Thus, in Shetland we used otter holts for population assessment, and for rivers and streams we developed a system based on spraints from otters with radio-transmitters, which could be recognized with the aid of a radio-isotope.

Numbers in Shetland

In Shetland we were commissioned by the Worldwide Fund for Nature (then the World Wildlife Fund) to estimate the total number of otters on the islands, and their distribution along the coasts. This survey was prompted by an outbreak in 1988 of distemper amongst common seals (*Phoca vitulina*); initially this was diagnosed as canine distemper (Osterhaus and Vedder 1988), to which otters are also known to be susceptible (Duplaix-Hall 1975). There was frequent contact between seals and otters in Shetland, and therefore considerable fear that otters could be very hard hit by the epidemic which killed thousands of seals.

For the Shetland survey we estimated the number of used otter holts in a 100 m strip along the coast, and subsequently we established otter numbers per holt. The Shetland landscape is open and relatively easily accessible, so it was possible to investigate all possible sites within the 100 m strip for the presence of holts, which we usually did with two observers. Many of the holts, in fact, were so conspicuous that one could see them from a long distance away. Only holts which were in active use at the time of the survey were counted; evidence for usage consisted of fresh spraints or tracks, smoothly worn lips of the entrances, matted and/or muddy vegetation, lack of cob-webs, and so on.

The statistical method for estimating the total number of holts has been described by Kruuk *et al.* (1989). It was developed originally for counting plains game in the Serengeti, East Africa, by Jolly (1969) and Norton-Griffiths (1973), and it has been found extremely useful in many different situations, keeping the variance of the final estimates to a minimum. I will briefly describe the method and how it works.

The total area to be censused is divided into a number of equal sections, in our case 5 km stretches of coast. Those sections are then grouped into a few 'strata' or habitats, based on either the expected density of holts (or animals), or on the observed densities after the counts have been made. For each stratum (h) the total number of holts (Y_h) is

$$\frac{N_h}{n_h} (\Sigma R_h)$$

in which N_h is the total number of 5 km sections in a stratum, n_h the number of sections which we censused, and ΣR_h is the sum of all holt counts (R_h) in each of those sections. The number of holts in Shetland is the sum of the numbers of holts in all strata.

The accuracy of this estimate, or the confidence limit, is derived as follows. First the population variance for each stratum is calculated:

$$\text{Var } (Y_h) = N_h \left[(N_h - n_h)/n_h \right] s^2$$

in which

$$s^2 = \frac{1}{n_h - 1} \left[\Sigma R_h^2 - \frac{1}{n_h} (\Sigma R_h)^2 \right].$$

The variance for the total number of holts in Shetland is the sum of the variances for each stratum, the standard error (s.e.) of the estimate is the square root of this variance, and the 95 per cent confidence limits of the estimate are at 1.96 s.e.

What makes this method especially useful is that it allows one to concentrate the census effort, the field work, on those habitats where there are many holts, and to sample those strata intensively whilst spending less effort in the low-density strata. This keeps the confidence limits of the total estimate as narrow as possible. Much of the secret of the success of the method, of course, rests on being able to stratify as efficiently as possible, using a minimum number of strata which cover the different density categories.

Some of the results from the Shetland survey have been mentioned in Chapter 2, showing the otters' preference for different habitats. Table 8.1 shows the total numbers of holts for the whole of the Shetland coast. Note that the confidence limits for individual strata are wide, but by combining the estimates and their variances the accuracy of the total estimate increases considerably.

The number of adult otters associated with these holts was estimated separately, in our intensive study area at Lunna and on a number of

TABLE 8.1 Numbers of otter holts along the coast of Shetland.

	Total no. 5 km sections (N_h)	% section counted (n_h)	Holts per section	Total no. holts (Y_h)	95 % confidence limits (% of Y)
Stratum					
I Cliff	89	21.3	1.79	159.3	51
II Agric	61	34.4	1.57	95.9	58
III Peat	40	55.0	13.27	530.9	17
IV Other	47	44.7	4.10	192.5	31
V Build	5	0	0	0	–
VI Islan	(133)*	35.3	7.77	206.6	27
Total	242+	35		1185.2	13

* For small islands (VI) the figure in brackets refers to 1-km sections, but for comparability the number of holts is calculated per 5 km.

small islands (Moorhouse 1988; Kruuk *et al.* 1989). We used the number of known resident adult females in each female group range and related that to numbers of used holts, because those animals had well-defined ranges of only a few kilometres. In a separate exercise, over larger areas, we estimated how many males and how many vagrants (non-residents) there were for every resident female. These estimates were made using the various systems of individual recognition of otters which we had developed, especially the throat patch differences and, where possible, coloured ear-tags.

We made three independent estimates of the total otter population in the main Shetland study area at Lunna. Andrew Moorhouse and I each estimated independently from our own observations each month how many individual otters we had recognized in each range, and we also pooled all our observations of otters with and without coloured ear-tags. From knowledge of how many ear-tags were swimming around Lunna in any one month we made monthly estimates of the population. The three methods gave remarkably similar results (Kruuk *et al.* 1989), which did not surprise us: we felt that we knew the population. The three different results were 16.5 ± 0.5 (Andrew), 15.1 ± 0.9 (me), and 16.0 ± 0.9 (ear-tags). We concluded that there were about 16 adult otters along the study coast (west side of Lunna Ness), or 1 per 1.2 km. Of those, nine were resident females, four resident males, and the rest were irregulars.

These estimates, and further ones made by Andrew on some of the smaller islands, were combined with the counts of numbers of used otter holts (Fig. 2.13, p. 47). This allowed us to infer the number of otters from numbers of holts, and on this we based our estimate for the total population of otters in Shetland. There were 0.331 resident females per holt, and the total number of adult otters was 1.83 times the number of resident females. With these data we estimated a total of 392 resident females in Shetland, and 718 adult otters in total (Kruuk *et al.* 1989).

Clearly there are several important assumptions behind this estimate, and these assumptions and their likely effects on the estimate have to be spelled out. The main ones were:

1. All holts in the 100 m strip were found in the surveys.

2. The sections surveyed were representative for the various habitat types.

3. The proportion of holts inside the 100 m strip in the survey was the same as in our intensive study area.

4. Individual otters in all habitats used similar numbers of holts.

5. The sex ratio and proportion of resident otters were similar everywhere.

Of these assumptions, (1) and (2) were probably quite realistic because of our intensive search and random selection of sections, with a high proportion surveyed (35 per cent). As to (3), there may have been somewhat more holts further inland in the cliff sections, and fewer in the agricultural areas, but these areas did not contribute much to the total, and if anything, the Lunna study area may have had somewhat more holts further inland than peaty coasts elsewhere. If, indeed, this was the case we would have slightly over-estimated the Shetland otter population. Numbers of holts used by otters, assumption (4), is an uncertainty which needs to be cleared up. However, even if, for instance, otters along cliff coasts used half as many holts as those in our peaty study area, this would cause our population estimate to increase by only 13 per cent, from 718 to 814. Finally, as to assumption (5) concerning sex ratio and percentage residents: possibly, there were more resident females in our Lunna study area than elsewhere on average (Chapter 2). If this were true, again our population estimate for Shetland would be too low. And as well as the possible sources of error discussed above, there must have been a few otters living further inland which were not covered by our survey.

Despite these cautions and possible objections, I think our assumptions were reasonably on the mark, and taking possible biases into account, as well as the sampling errors and the statistical error of converting holt numbers into otter numbers, I am confident that the number of otters in Shetland in 1988 was somewhere between 700 and 900 animals, or 0.5 to 0.7 per kilometre of coast (Kruuk et al. 1989). This is close to the figure of 0.8 otters per kilometre of coast on the Shetland island of Fetlar (Baker et al. 1981), and also other naturalists and conservation management staff who know the coasts very well feel that this figure is about right.

In September 1993 the Shetland survey was repeated, in order to assess the effects of the oil spill caused by the stranding of the tanker *Braer*. At the time of writing the results have not yet been fully analysed, but it appeared that there were fewer otter holts used close to the site of the wreck at the south end of Shetland, and somewhat more in all other parts of the islands. Thus, the Shetland otter population appears to be doing well over these years.

Numbers in freshwater areas

The method which we used for assessing otter numbers and utilization of inland streams and rivers was different from that used in Shetland (Kruuk et al. 1993). It was based on otters fitted out with radio-transmitters, and injected with a harmless radionuclide (Zinc-65) which enabled us to recognize their spraints with a scintillation counter. In principle, we established the home range of 'focal animals', the ones with transmitters, and we assumed that other otters in those ranges would be as likely, on average, to spraint along any one bank at any one time, as the focal animals. This assumption was based on the observation in Shetland that there were no significant differences in sprainting rates and seasonality between otters of different sex or social status (Kruuk 1992). Then, together, the proportion of spraints with the radionuclide, and the proportion of time spent in the area by the focal otter allowed us to make an estimate of time spent in that area by all otters present over the study period.

To estimate the usage of a relatively large stretch of river by otters, we calculated the proportion of 'active' time which the focal animal spent in a particular section. The animals were almost exclusively nocturnal, and to overcome the problem of radio-tracking periods of varying length during different nights, we chose as the unit of time spent the 'otter night', one activity period (that is one night, of about 4 to 5 hours) spent by one otter in the area. For instance, if an otter was tracked over

2 hours during one night, and it spent half an hour in the Esset Burn and the rest of the time in the River Don, then 0.25 otter nights would be allocated to the Esset, 0.75 otter nights to the Don. When we ended the radio-tracking of one particular animal after several months (in some cases more than a year), the usage of a section by that individual was expressed in otter nights per year. Then, taking into account the proportion of spraints left by other otters, the total number of otter nights per year spent by all visiting otters could be calculated.

As an example, one male otter (8.0 kg) was the focal animal from November 1989 until June 1990. He spent most time along the River Dee (69 of 125 nights radio-tracking, or 55 per cent), but foraged along the Sheeoch Burn, a small tributary of the Dee, during 47 nights (38 per cent). He used 11.6 km of the Sheeoch (or 5.1 ha water), and spent the equivalent of $47/125 \times 365 = 138.7$ nights/year there. Of 700 spraints collected along the Sheeoch during the observation period, 78 per cent contained the radionuclide (mean 77.2 ± 5.4 per cent s.d. over five collecting periods), suggesting a total number of otter nights/year along the Sheeoch Burn of $100/78 \times 138.7 = 177.8$. This figure could be converted to a mean nightly otter biomass of 4.62 kg for the whole of the Sheeoch, or 0.091 g m^{-2}.

The size of otters' home ranges have been discussed in Chapter 2 (p. 58); they varied between about 20 and 80 km of stream in the north-east of Scotland, at least for those animals which we could follow over a sufficient period. The overall, median value of density of otters in the Rivers Dee and Don and their tributaries was 1 otter per 15.1 km of stream, but it varied between 1 per 3 and 1 per 80 km of stream, or 1 otter per 2–50 ha of water. As shown in Chapter 2, this variation in density could be related to the width of streams (Kruuk et al. 1993).

Comparison of the freshwater data with those on otters along coasts in Shetland suggests, at first sight, that Shetland otters live at much higher densities. However, if otter numbers are expressed per area of water instead of length of bank, the discrepancy largely disappears. If we assume the strip of water used by coastal otters to be 80 m wide (enclosing 98 per cent of all otter dives)(Kruuk and Moorhouse 1991), the estimate for good otter habitat in Shetland is one animal per 10 ha water, the same order of magnitude as densities in streams and lakes.

At present we do not yet have reliable estimates for total otter populations over larger areas of mainland Scotland, but with figures such as the above, and information on lengths and widths of rivers and tributaries, it should soon be possible to arrive at a useful overall estimate.

AGE STRUCTURE AND LIFE EXPECTANCY

Several methods have been used to determine the age of individual mammals, including measurements of eye lenses, baculum and various other bones, and tooth wear and incremental rings in teeth, methods usefully reviewed by Morris (1972). Of these, the counting of annual incremental rings in teeth is the most accurate for animals such as carnivores, and for otters the method was developed by Heggberget (1984). She used the incremental rings in the cementum at the tip of the root of one of the small incisors or of the much larger canines (Fig. 8.1). To do this a tooth is decalcified, sectioned with a freezing microtome, sections are stained, then rings can be counted with 64× magnification.

Previous to our observations from Shetland, Stubbe (1969) in Germany (recognizing adults and subadults from size) and Heggberget (1984) in Norway collected data on the age of otters which had mostly been shot by hunters. They found high percentages of juveniles, but such samples are likely to be somewhat biased towards younger animals,

FIG. 8.1 Section through the root of an otter's canine tooth, showing layering in the dentine in a 3-year-old animal.

as they are more easily approached and shot. Similarly, most information on *Lutra canadensis* comes from harvested populations which showed a relatively young age structure (Stephenson 1977; Tabor and Wight 1977; O'Connor and Nielsen 1981). We had little idea about what to expect from the observations on otter ages in Shetland, therefore, apart from an intuitive picture of rather long-lived animals—which proved to be quite wrong.

Otters can live to a respectable old age, up to 15 years in captivity (Chanin 1985); in our enclosure we have a female of 12 years old which is still going strong at the time of writing, and the same-sized *L. canadensis* has a reported longevity in captivity of 25 years (Melquist and Dronkert 1987). In the wild, we found the oldest animal, a female, killed on the road in Scotland at the age of 15 years. All the more surprising, therefore, that in general the age structure of otter populations we looked at is very young: a wild otter has a very short life expectancy (Kruuk and Conroy 1991). The following figures on mortality of otters in Shetland include data from a smaller sample which were published earlier (Kruuk and Conroy 1991); the conclusions here are similar.

In Shetland we ourselves found dead otters on the roads, or along the shores where the animals lived (Fig. 8.6), but most otter carcasses were found by Shetland people. The bodies were collected from all over the islands by the then Nature Conservancy Council (now Scottish Natural Heritage) in Lerwick, and kept in a freezer until we collected and analysed them. Their ages are shown in Fig. 8.2.

Most of the animals had died young, before they were 3 years old. Probably this phenomenon was even stronger in reality than as shown in Fig. 8.2, because it was likely that young cubs, which spent almost all their time under ground, were never found if they died. However, the picture was somewhat different when we expressed annual mortality as a percentage of each year class. For this, we assumed that the sample of dead animals was representative of the whole population, that is we treated our sample as a population of 113 otters, of which 21 died before the age of 1 year, and so on. If then we plot annual mortality against age (Fig. 8.3), a clear picture emerges of linearly increasing probability of death with age. This is unusual in mammals: most species show a pattern of mortality which is high when very young, then lower in adult age, sometimes high again at a ripe old age (Caughley 1977). In a (very high density) population of badgers there was a large mortality amongst cubs and 1-year-olds, then a fairly high death rate amongst adults (mean 21.8 per cent), without any specific trend with age (Cheeseman *et al.* 1987).

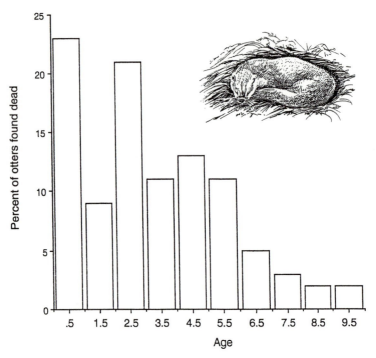

FIG. 8.2 Age distribution of otters found dead in Shetland, 1984–8. Data from 113 animals. Age 1.5 = animals between 1 and 2 years old, etc.

As explained above, we should be wary of mortality figures of animals less than 1 year old, but there is no reason to assume that the probability of finding a 3-year-old animal is different from that of finding a 10 year old. To look at the pattern of otter survival, therefore, it is useful to start with animals of 1 year old, at the beginning of independence of individuals, and express the probability of mortality before a given age (Fig. 8.4). In Shetland more than half of those recruits to the adult population die well before they are 4.5 years old: their mean life expectancy at the age of 1 is 3.14 years (calculated as in Southwood (1978), with a sample size N of 94 adults). Such high mortality is not unusual amongst carnivores of this size; for instance, badgers show a similar trend, with few animals in a high-density population reaching ages beyond 6 years (Cheeseman *et al.* 1987).

There are small differences in the life expectancy of otters depending on whether calculations are based on animals which died violently (almost all on roads) or non-violently, that is 'naturally': from the road-

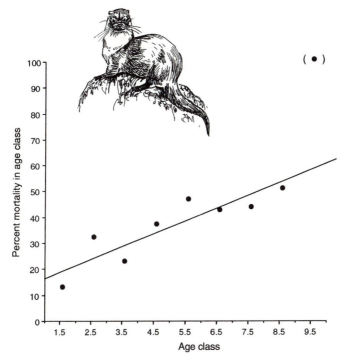

(•)

FIG. 8.3 Mortality of Shetland otters at different ages, showing the percentages of those animals which were alive at a given age and which died in the subsequent year. $r = 0.93$ ($p < 0.001$). Regression excluding the oldest year class: $y = 11.93 + 4.65x$. (From Kruuk and Conroy 1991.)

kill sample I calculated a life expectancy of 2.86 years ($N = 50$), from the natural deaths this was 3.57 years ($N = 44$). Obviously the true figure lies somewhere in between, and I will argue below that it will be closer to 3.57 years, the expectancy as based on non-violent deaths. Females can expect to live slightly longer, on average 3.22 years ($N = 54$) as compared with 3.02 years for males ($N = 40$, all samples).

This low life expectancy for otters appears to be a common feature also in other areas. For instance, we analysed a sample of 145 otters of 1 year or over, collected between 1982 and 1990 mostly from mainland Scotland. This also included a few animals from England; almost all were killed on roads, collected by many different people and given to us or to the Nature Conservancy Council. In that sample the mean life expectancy was 2.73 years, as compared with 2.86 years for the road-kills in Shetland.

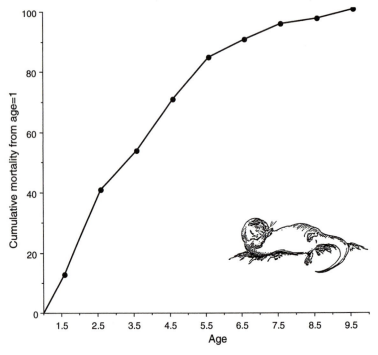

F IG. 8.4 Cumulative mortality of otters in Shetland, showing the percentage of those otters which were alive at 1 year old, and which had died by a given age. Both sexes, all causes of death ($n = 113$ otters).

MORTALITY

Proximate causes of death, and body condition

In a species such as the otter, where animals usually die long before achieving their potential lifespan, the causes of death are very important for the understanding of factors which limit populations. If the animals lived to ripe old age it would matter little whether they died under the wheels of a vehicle, from kidney stones, or through starvation.

Many different diseases and other ailments have been mentioned in the literature on otters (summaries in Harris (1968), and for this and other species in Melquist and Dronkert (1987)), but few quantitative data are available on their importance as causes of mortality. We first got some data on causes of death in our study of otters in Shetland (Kruuk and Conroy 1991).

A major problem in quantifying such information is the fact that the various causes of death have different probabilities of being detected. For instance, any fatalities associated with the activities of man are noticed immediately, but a quiet death in a remote spot will escape attention. Previous authors have emphasized the slaughter of otters on roads (Fig. 8.5), in fyke nets set for eels, in lobster creels, by hunts, and various other means (Wijngaarden and van de Peppel 1970; Chanin and Jefferies 1978; Stubbe 1980; Twelves 1983; Jefferies *et al.* 1984; Chanin 1985; Mason and Macdonald 1986; and others), but no data have been presented enabling the evaluation of these observations against other causes of mortality. In Shetland the same problem applied, but to a somewhat lesser extent: the habitat is much more open and accessible than in most other otter areas, and otters dying in remote places are more likely to be found than, for instance, in mainland Britain. In the end, however, it was still impossible to ascertain the magnitude of the detection bias, and it must have been large.

Of 113 otter carcasses from Shetland, 55 (49 per cent) were killed on roads (Kruuk and Conroy 1991). However, it was concluded that, with the large bias favouring road-kills, traffic mortality was a relatively small factor of unknown magnitude. Other kinds of violent deaths included one otter drowned in a lobster creel, and four which had been bitten to death, either by dogs or other otters. A total of 46 per cent died from non-violent causes (Fig. 8.6), with symptoms which, on autopsy by a veterinary scientist, included haemorrhaging in stomach or intestines (8 per cent of all carcasses), liver abnormalities (lesions, hepatic neoplasm) (2 per cent), pneumonia (1 per cent), and poisoning

FIG. 8.5 Traffic sign in Shetland (near Sullom Voe). Many otters die on roads.

FIG. 8.6 Non-violent death of a 5-year old female otter in Shetland.

by oil or paint (2 per cent). However, in most of these non-violent deaths
no immediate cause could be diagnosed, either because there were no
significant abnormalities (9 per cent), or because the carcass was too
decomposed (25 per cent). Considering only the 24 bodies of non-violent
deaths which could be analysed, gastro-intestinal haemorrhaging occurred in 37 per cent: that is a symptom which may be associated with
starvation.

The immediate cause of mortality varies greatly, therefore. One characteristic of the non-violent deaths was their poor overall body condition: the animals looked worn-out and thin, possibly because of
starvation.

To express body condition more quantitatively, we weighed all of
them and calculated a 'condition index' K, to take into account the substantial differences in overall body length between otters. This is the
same condition index as is used for fish (Le Cren 1951). It uses the fact
that in most animals the relationship between average weight and
average body length can be expressed as

$$W = aL^n,$$

in which a and n are constants, determined by the shape of the body,
amongst other things. In a sample of 25 road-kills (therefore presumably 'healthy' otters) we found that, on average, for females $a = 5.02$,
$n = 2.33$, and for males $a = 5.87$, $n = 2.39$, with weight in kilograms and

length (straight line from nose to tip of the tail) in metres. So as weight-for-length, males are relatively heavier than females, on average.

The condition index for each individual can now be calculated as

$$K = W / 5.02L^{2.33}$$

for females, and

$$K = W / 5.87L^{2.39}$$

for males. If, in the above sample of road-kills, we use the calculated constants we get an average $K = 1.00$. This, therefore, is the average, normal, healthy otter. An animal which is under weight will have a K which may be as low as 0.5, and a really heavy otter may have a $K = 1.4$ or more.

Figure 8.7 shows the differences in condition indices K for otters which died in Shetland, with a mean of 0.77 ($\pm$ 0.15 standard deviation, s.d.) for the non-violent deaths, and 1.08 ($\pm$ 0.15 s.d.) for those which were killed violently (Kruuk and Conroy 1991), which is a highly significant difference.

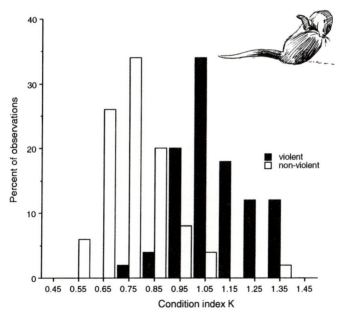

FIG. 8.7 Body condition K of Shetland otters which died a violent (48 animals) or non-violent death (49 animals). K = 1 is 'normal', K < 1 is underweight, etc. 'Violent' deaths, mostly traffic victims, were in significantly better condition ($\chi^2 = 68.2$, d.f. = 5, $p < 0.001$). (After Kruuk and Conroy 1991.)

More recently we collected data on carcasses of otters which died in mainland Scotland. For instance, of 76 bodies which were sent to us or collected by ourselves in 1991–2 for the purpose of analyses for pollutants, 65 (86 per cent) were of otters which had died violently, almost all on roads. The mean condition index of the road-kills was 1.04, compared with a mean for the non-violent deaths of 0.78. Clearly the figures for the mainland and for Shetland are very similar, except that on Shetland we found more of the non-violent, 'natural' deaths.

Effects of pollution

The evidence suggests that otter populations in England and several other countries in Europe have been decreasing since the nineteenth century, but that the dramatic, sharp, and in many places ultimate decline started in the 1950s (summaries by Mason and Macdonald (1986) and Jefferies (1989)). There is little doubt in anyone's mind that the cause of this final decline and regional extinction is pollution. The disappearance of otters from many waters all over Europe coincided with a massive increase in the use of organochlorines as insecticides in agriculture, and several of these substances have been found in damaging concentrations in the tissues of dead otters. Many other fish predators, birds of prey, and others have suffered similar fates, demonstrably due to pesticides. Many of the compounds which were held responsible for wildlife deaths have now been taken out of use and many of the species which were affected by them have recovered—however, otter populations appear to be notably slow in returning, and in some areas continue to decline.

We still know relatively little about which compounds are responsible for the demise of the otter. The possibilities are many (Mason and Macdonald 1986; Mason 1989), it is mostly unknown at what level they may affect otter survival, and it is obviously impossible to do critical experiments. Here I want to discuss observations on the present-day role of some of the most commonly blamed compounds in otter mortality in areas of Scotland.

If substances are having toxic effects on individual otters, one could expect to see a negative correlation between physical condition and concentration of the compound. Then one would also expect either an accumulation of toxic substances in animals with age, with compounds gradually increasing their impact or perhaps becoming effective beyond some threshold concentration, or there would be sudden deaths with high concentrations in the body. If animals excrete suspect compounds as fast as they absorb them, then it is unlikely that such substances will directly affect mortality—though they may affect prey populations.

There has been a considerable amount of analysis of otter carcasses for different pollutants, especially in Britain by Mason (1989), but also in other countries, for example in Sweden by Olsson *et al.* (1981) and The Netherlands by Broekhuizen (1989). There is little information on either age or body condition of the animals, but those previous studies provide a good overview of the concentrations in which the compounds are found. We sent samples first from 113 otters from Shetland, then from 116 otters from different parts of Scotland for analysis to the chemical laboratories of the Institute of Terrestrial Ecology in Monks Wood, near Cambridge. The results were complicated, as could be expected, but they enabled us to draw some tentative conclusions (Kruuk and Conroy 1991; Kruuk *et al.* 1993).

The samples were analysed for polychlorinated biphenyls (PCB), for the organochlorines DDE (the breakdown product of DDT), dieldrin (HEOD), lindane (BHC), and for mercury (Hg). Of the first set of samples, from Shetland, some carcasses were also analysed for cadmium (Cd), lead (Pb), and selenium (Se).

Almost all of these, with the exception of PCB and mercury, occurred in concentrations well below the levels known to have significant lethal or sub-lethal effects in mammals or birds (for example Jefferies 1969; Robinson 1969; Bunyan *et al.* 1975; Wren *et al.* 1988). With the exception of mercury, none was correlated with the age of the otter, so they were not likely to accumulate and produce complications later, at least not at the rate at which they were consumed at the time of collection. Also, none of the compounds occurred in concentrations which were correlated with the body condition K. Of all the substances analysed, therefore, only PCBs and mercury were candidates for causing possible deleterious effects.

PCBs are suspected of playing a major role in the demise of otters in Britain and elsewhere in Europe (Mason 1989; Mason and Macdonald 1993; Mason and Madson 1993), and as they may be distributed as aerial or aquatic (industrial) pollutants they could have an impact almost anywhere. Mason (1989) suggested that their concentrations in otter tissues from different countries are related to the status of otter populations, and it has been shown in the laboratory that PCBs cause failure of reproduction in female mink (at concentrations in liver lipids of 50 ppm) (Jensen *et al.* 1977). Mason (1989) hypothesizes that otter populations with PCB values of 50 ppm (lipid in liver) or over should be declining. From this, I predicted that the apparently dense and healthy population of Shetland otters would have PCB values well below 50 ppm.

However, in our results the PCB values were relatively very high, with a mean of more than 210 ppm lipid (5.5 ppm wet weight; $n = 73$) in the liver. Non-violent deaths had significantly higher concentrations than the ones which were killed by traffic (356 ppm. lipid weight, $n = 34$, but even in those 'healthy' road kills the mean value was almost 80 ppm ($n = 39$). There was no correlation between PCBs and age, no evidence of accumulation, perhaps because otters are able to metabolize or excrete this pollutant. Our results do not support the Mason hypothesis.

The story of mercury in Shetland otters is even more interesting. Mercury does accumulate with age (Fig. 8.8), and in the livers of otters of 5 years old or more, we found concentrations of up to 60 ppm (dry weight). In the American otter *Lutra canadensis* some unpleasant, but useful, experiments by O'Connor and Nielsen (1981) showed that a mercury level of 33 ppm wet weight in the liver (which is about 100 ppm dry weight) indicates a lethal level of mercury intake.

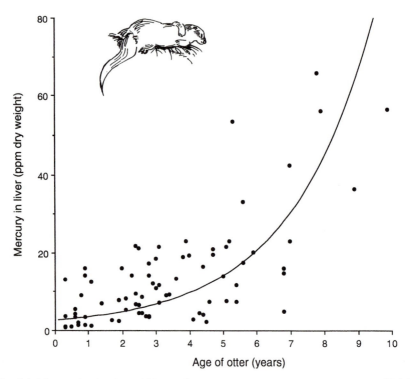

FIG. 8.8 Mercury in the liver of Shetland otters, in relation to their age. Mercury (Hg) in ppm dry weight. Spearman correlation $r_s = 0.63, p < 0.001$. Regression: ppm Hg = $1.86 \times 10^{0.17\text{age}}$. $r^2 = 0.32$. (From Kruuk and Conroy 1991.)

Thus, if *L. lutra* responds the same as the American river otter, then the concentrations of mercury which we found in the Shetland animals were very high and could have had sublethal effects. The accumulation with age could cause mercury to be a significant contributor to overall mortality. It could also explain some aspects of the increase in mortality with age, as shown on p. 215. There is much more to be studied here; amongst others, there is the interesting situation that mercury occurs naturally in relatively high concentrations in the seas around Shetland, probably derived from submarine volcanic activity (Carr *et al.* 1974), and causing high mercury levels in marine fish (Davies 1981; Kruuk and Conroy 1991). But mercury is also an aerial pollutant (Piotrowski and Coleman 1980), and Shetland may be affected by industrial pollution from elsewhere.

Many sea mammals absorb mercury in relatively high quantities, but they are able to use selenium to counteract its toxicity (Koeman *et al.* 1975; Smith and Armstrong 1978; Reijnders 1980). Those species accumulate and store selenium at the same rate as mercury, and on analysis the above authors found a near-perfect correlation between the two elements in species such as seals and dolphins. Wren (1984) suggested a similar relationship in semiaquatic freshwater mammals in Canada. However, in our Shetland otters there was no correlation between mercury and selenium ($r = -0.07$, not significant), which suggests that in these animals no such selenium mechanism evolved to cope with the mercury problem (Kruuk and Conroy 1991).

In the second set of samples, of otter carcasses which were collected from many parts of Scotland including Shetland and Orkney, there was further evidence for some of the main points made in Shetland. Again, compounds such as dieldrin, DDE, and lindane were nowhere present at biologically significant levels (Table 8.2), and again, the only substance which correlated with age of the otters was mercury.

In this sample there was a significant negative correlation between body condition K with both PCBs and mercury ($r = -0.45$ and -0.29 respectively, both highly significant): otters in bad condition contained more PCB and mercury. It is possible that low body condition was at least partly caused by the presence of the toxic compounds, but alternatively it is also possible that in otters in bad condition body fat is mobilized, which would free various compounds present in these lipids and concentrate them in remaining lipid stores, for example in the liver. From our samples there was no way of deciding between those options.

Comparing samples from the various areas in Scotland in this second study, we found that levels of mercury and PCBs in the otters from Shetland were comparable with those in the first. However, in most

TABLE 8.2 Summary of values of the main compounds analysed in 116 otter carcasses from mainland Scotland, Orkney and Shetland, in parts per million (ppm).

Compound	N	Mean	Std dev.	Minimum	Maximum
Merc (wet)	115	4.04	3.01	0.11	13.43
Merc (dry)	115	13.25	10.20	0.33	44.71
DDE (wet)	116	0.25	0.42	0	2.81
DDE (lipid)	116	9.64	17.68	0	116.78
HEOD (wet)	116	0.10	0.05	0.02	0.28
HEOD (lipid)	116	3.58	2.44	0.67	14.00
PCB (all; wet)	116	1.49	2.12	0	14.40
PCB (all; lip)	116	57.35	91.76	0	595.92

other places otters had even more mercury in their livers than in Shetland. PCBs, on the other hand, were higher in Shetland than anywhere else. The mean PCB concentration in Shetland was 140 ppm (lipid), the median 72 ppm, with 25 per cent of otters having levels above 152 ppm (compared with 50 ppm causing reproductive failure in mink).

Especially striking was the observation that several lactating females (that is successfully reproducing animals), which had been killed on roads on the Scottish mainland, contained high levels of PCB. One female had 1096 ppm (lipid) in her liver, more than 20 times the concentration which causes reproductive failure in laboratory mink. One of the obvious conclusions from all this is that the toxicity of PCBs for otters is nowhere near as high as for mink, and one should not indulge in the common practice of using the effects on one species as a yardstick for another.

To evaluate the chemical concentrations in wild otters we should know details of the status of the populations: good, poor, declining, increasing, or whatever. We do not often have this information, but we found the Shetland population to be dense and healthy. In the other areas of Scotland from which we received samples there may be fewer otters than in Shetland, but there do not appear to be obvious trends in otter numbers, either up or down.

If these assessments are correct, then the data support the hypothesis that PCBs have little effect on otter populations at present, at least not in Scotland. Mercury may make a contribution to otter mortality which increases with age, but does not necessarily affect population density if recruitment can make up for it. Both PCBs and mercury can be aerial as well as aquatic pollutants, with mercury occurring naturally and as a man-made contribution to the environment, from industry and agriculture.

An important point is that the role of pollutants in otter mortality will have to be considered jointly with other effects, such as food availability and recruitment to numbers. Pollution can act also indirectly, by affecting numbers of prey. I will further discuss the effects of pollutants in Chapter 10.

The role of food availability

The effects of lack of food, of starvation, as a cause of mortality, are complicated because they act together with other environmental stresses, and it is rare to find any animal dying just of starvation. If food shortage were a major factor causing mortality, one could expect more otters to die (of whatever proximate cause) at times of low food availability, and we should see low body condition associated with that.

Mortality of otters other than by violent death is, indeed, highly seasonal (Kruuk et al. 1987; Kruuk and Conroy 1991). On Shetland we found most 'natural' deaths in spring, with 60 per cent of all natural mortality occurring between March and June. This was a highly significant seasonality in sharp contrast with the even distribution of violent deaths throughout the year (Fig. 8.9). Similarly we reported that on the Scottish mainland, 42 per cent of all non-violent mortality ($n = 19$) occurred in April alone, more than in any other month, whilst violent mortality ($n = 73$) was evenly distributed over the months (12 per cent in April), a highly significant difference (Kruuk et al. 1993). Such seasonality is parallelled by the availability of the important prey species both in the sea and in fresh water (see Chapter 5; Kruuk et al. 1987, 1988, 1993). The low prey availability occurs at a time of year when otters' metabolic requirements are high, due to low water temperatures (Chapter 7).

The expected seasonality in body condition of otters, as expressed by the index K, was not evident in our data: otters which are killed on roads or by other violent means show the same relative body weight throughout the year (Kruuk and Conroy 1991). Nevertheless when an animal dies 'naturally' in spring, its body weight is significantly lower. The most likely explanation for this is that otters have only small fat reserves at any time of year, and that when an animal starts losing condition, that is starts using those reserves, it is likely to die very soon.

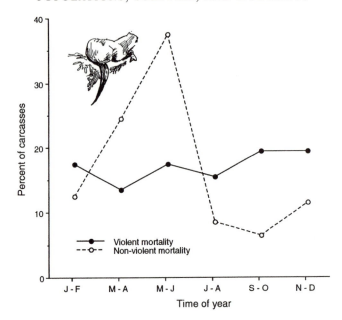

FIG. 8.9 Mortality of Shetland otters at different times of year, as percentages of all violent and non-violent deaths for each 2-month period. Violent deaths: 53 otters, $\chi^2 = 0.78$, d.f. = 5, not significant. Non-violent deaths: 49 otters, $\chi^2 = 20.1$, d.f. = 5, $p < 0.01$. (After Kruuk and Conroy 1991.)

There was evidence from the freshwater areas we studied in north-east Scotland that otters may eat a large proportion of the available prey population (salmonids), and that during a time of low prey availability they increasingly switched to suboptimal types of prey such as carrion, birds, and mammals (Kruuk *et al.* 1993), as they did in Shetland. Otter numbers in rivers and streams were closely correlated with fish productivity (p. 153). From all this evidence it seemed likely, therefore, that food shortage was a major ultimate factor causing mortality of otters in both the sea and fresh water.

RECRUITMENT

Both sexes usually reach sexual maturity in their second year of life (Corbet and Harris 1991), as in the American river otter. In this latter species there is considerable variation in the percentage pregnancy of 2-year-old females, whilst almost all older females become pregnant (Toweill and Tabor 1982; Stenson 1985 in Melquist and Dronkert 1987). Such variation is also likely to occur in *Lutra lutra*, but it has not yet been

documented, and the data will be difficult to collect for separate populations because, unlike *L. canadensis*, *L. lutra* is not commercially
exploited.

Litter sizes of Eurasian otters are small, reaching four (Fig. 8.10), but
usually keeping well below that number. In Shetland, the mean number
of cubs per litter was 1.86 ± 0.61 (s.d.; $n = 102$ litters) along the Yell
Sound between 1980 and 1985, and in our main study area at Lunna it
was 1.64 ± 0.73 ($n = 28$; Kruuk *et al.* 1991). Taking sizes of litters when
they were first observed in Lunna (cubs about 2 months old), in 28 observations 13 (46 per cent) consisted of one cub only, 13 of two cubs, and
in the remaining two there were three and four cubs each. Also elsewhere in coastal areas litters were small, with means of 1.95 cubs in
Norway, and 1.55 along the Scottish west coast (Kruuk *et al.* 1987). In
inland areas family sizes are larger: 2.8 in The Netherlands (when
otters still occurred there), 2.3 in East Germany, 2.4 in Poland, and 2.5
in inland areas of the United Kingdom (Wijngaarden and van de Peppel
1970; Stubbe 1980; Wlodek 1980; Mason and Macdonald 1986).

Numbers of cubs born per adult female per year are more difficult to
determine than litter sizes. In Shetland (Lunna) over 5 years a mean of
9.2 cubs were found per annum for a total of nine simultaneously resident females (1.02 cubs per adult female per year). To give some examples of individual females and their reproductive successes: F1 had one
cub in 1984 which she lost, then one cub in 1985, and she disappeared in
1986. F7 reared one cub in 1984, none in 1985, two in 1986, and she disappeared in 1987. Her 1984 cub (F16) stayed in the same range, reared

FIG. 8.10 The largest family in the study area: female with four cubs, asleep on bladder
wrack (*Fucus vesiculosis*).

her first cub in 1986, and probably none in 1987. F33 was quite old when we first caught her (judging from tooth wear), and she had no cubs in 1985, 1986, and 1987, then disappeared. F11 had litters of two cubs in both 1985 and 1986; F44 had one cub in 1986, three in 1987 (of which she abandoned one; Fig. 8.11), and of these last two animals the subsequent histories are not known.

Of course, these data do not reflect actual pregnancy rates of females, because litters may have been lost pre- or even post-natally (inside the natal holt) without being registered. It is known that females may abort, for instance in the wild as a consequence of bacterial infection (Weber and Roberts 1990).

Reproduction of otters in Britain is remarkably unseasonal, and cubs may be born in any month, a fact which has been noted by many authors (Stephens 1957; Harris 1968). On the European continent births are more common in spring and summer (Erlinge 1967; Wijngaarden and van de Peppel 1970; Danilov and Tumanov 1975; Reuther 1980), but in general, in many places otters are far less synchronized in their breeding season than any other carnivore (Corbet and Harris 1991). Surprisingly though, we found otters in Shetland with a distinct breeding season (Kruuk *et al.* 1987): calculating 2 to 2.5 months back from the dates of our first observations of otter cubs during their first weeks out of the natal holt, there appeared to be a clear birth peak around June (Fig. 8.12). From similar observations something of an aggregation of births was found also for the coasts of north-west Scotland, but far less pronounced (Fig. 8.12).

FIG. 8.11 Two-month-old cub, abandoned by its mother.

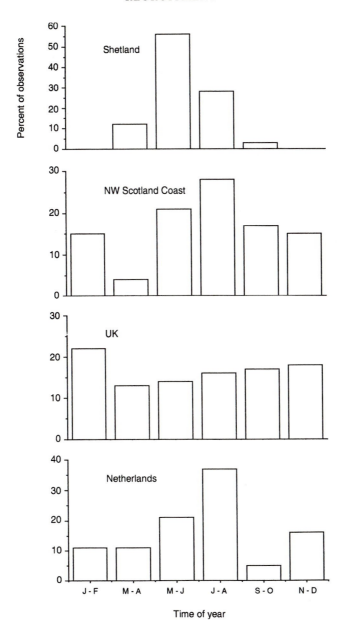

FIG. 8.12 Breeding seasons of otters in four different areas. Data for Shetland from Kruuk *et al.* (1987), for north-west Scotland from J. Rowbottom (personal communication), for the whole of Britain from Harris (1968), for the Netherlands from Wijngaarden and Peppel (1970). Statistical significance, litters May–August versus rest of year: Shetland 34 observations, $\chi^2 = 37.3$, $p < 0.001$; north-west Scotland 47 observations, $\chi^2 = 5.1$, $p < 0.05$; Britain 168 observations, $\chi^2 = 0.2$, not significant; Netherlands 19 observations, $\chi^2 = 5.2$, $p < 0.05$.

Also when we recorded mere presence or absence of cubs (irrespective of size or age) accompanying adult otters, at different times of year in Shetland, a very clear seasonality was evident (Fig. 8.13). The data demonstrate a more or less steady percentage of family parties in the population from October onwards, with families splitting up around April.

The difference in the timing of breeding between Shetland and elsewhere is striking. This is not likely to be due to a more pronounced seasonality in climate in Shetland; if anything the differences between winter and summer are less there (Berry and Johnston 1980; Kruuk *et al.* 1987; Fig. 7.1 on p. 185), tempered by the seas around the islands. Having seen the relationship between otter mortality and food availability, we therefore postulated a similar hypothesis to explain the seasonality in breeding. This suggests that the breeding season is timed in such a way that the greatest energy requirement of the female coincides with the greatest availability of fish.

The energetic needs of female mammals are highest at the time of lactation (Randolph *et al.* 1977; Widdowson 1981; Loudon and Racey 1987). In fact, it is estimated that, in general, the energetic requirements of a female during one day at peak lactation exceed the total en-

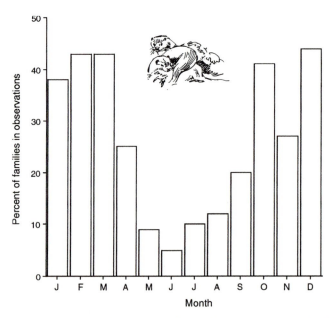

FIG. 8.13 Otter families seen in Lunna Ness, Shetland, as proportion of all sightings per month, 1983–7. Total 1505 observations.

ergetic contents of a litter at birth (Oftedal and Gittleman 1989). In the case of otters, the cubs are provisioned with fish starting at the age of about 2 months, and they are weaned slowly over a period of several months. We may expect peak lactation to occur between 1 and 3 months after birth, in Shetland between July and September.

We registered, indeed, an up to 10-fold increase in August in the number of eelpout in our study area (p. 144). That was one of the most important prey species overall there and by far the most important during summer, and in the areas where most females have their cubs, near the shallows (p. 138). This supported our hypothesis, but the relationship between seasonalities of otter breeding and fish availability may go even further.

We also established that during autumn and winter otter diet includes more relatively large fish, again related to availability (Chapters 4, 5; pp. 110). Female otters need these relatively large prey to carry to their offspring: they can take only one fish at a time, and often cubs are waiting ashore whilst the mothers fish (Fig. 8.14). It does not pay to carry small prey over a long distance to the cubs, therefore large fish are needed (Chapter 3, p. 90). In other words, cubs are growing up and are being provisioned by their mother just at the time when more large fish are available, making for increased efficiency of the female's provisioning behaviour.

Fig. 8.14 Large cub (estimated 8 months old), provisioned by mother.

If the seasonal increase in food availability is an immediate causal factor underlying the timing of otter breeding, it is possible that the actual amount of prey in the territory during the peak of lactation has an effect on recruitment, on the number of cubs raised per female. This suggestion is interesting, especially since the number of fish caught in our traps during July and August over several years varied a great deal. To test the prediction we compared the numbers of otter families and cubs in our Shetland study area, observed during the winter months between 1984 and 1989, with the numbers of potential otter prey which we caught in our fish-traps during the preceding month of August (Kruuk *et al.* 1991).

There were very strong correlations between numbers of cubs in the study area and numbers of prey (Fig. 8.15), numbers of families and numbers of prey ($r_s = 0.97$, $p < 0.05$), and a positive but non-significant correlation between numbers of cubs and biomass of prey ($r_s = 0.70$). The number of fish during July and August was not correlated with fish

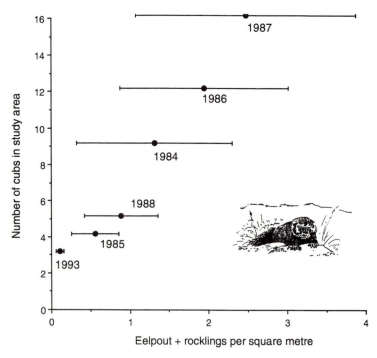

FIG. 8.15 Numbers of otter cubs along 20 km of coast at Lunna Ness, Shetland, as observed during autumn and winter 1984–8 and in 1993, related to mean density (± 95 per cent confidence limits) of eelpout and rockling in previous July–August (estimated from fish trapping). $r = 0.99$, $r^2 = 0.96$; $p < 0.001$. Includes data from Kruuk *et al.* (1991).

numbers at other times of year, so if there were a causal relationship between August fish and otter cubs this was not because cub numbers were affected by fish at some other time.

In 1993 we trapped fish again in August, and we counted cubs in November and December (to use the Lunna Ness study area as a control for observations of the effects of the *Braer* oil spill). In the study area fish populations were the lowest we ever recorded and, as predicted, so were the numbers of otter cubs (Fig. 8.15).

It is not clear exactly what happens to bring about this correlation between summer prey population and winter cub numbers. One explanation could be that cubs die during their first 2 months of life in the holt, before they emerge and can be counted. It was not possible to verify this. However, on several occasions we have observed deliberate abandonment of individual cubs by mothers, immediately after the family leaves the natal holt (Fig. 8.11). The following describes one of those incidents:

July 1986, Boatsroom Voe. Just by chance I see one of our females, Mrs Fitz, carry a small cub down the hill [Fig. 8.16]. She comes from the natal holt which she has used over the last couple of months, and I had been expecting her to emerge with her offspring for some time now. The natal holt is about 300 m inland, quite high up, and she carries and drags the little one all the way down, to a large holt on the shoreline which has been out of use for several months. On arrival there is another cub sitting in one of the entrances, so what I am watching is not her first trip. She and the two cubs disappear undergound, and for two hours nothing more happens. By then it is afternoon, and I see her emerge again, going down to the shore, alone. She splashes about in the partly exposed algae, knotted wrack, and comes up with a large bunch of the weeds in her mouth, which she takes into the nearby holt. Ten minutes later she does the same, obviously nest building. Another hour passes before she emerges again, this time going uphill, back to the natal holt. Soon she is back at the coastal holt, dragging, carrying a third cub. Then all settles down, with the family inside the holt, out of sight; from a distance, we quietly keep an eye on it.

Later, the evening of the same day, the otters are on the move again. Mrs Fitz emerges, three cubs follow, all going down the bank, scurrying around on the sea weed. Mrs Fitz dives in, comes up, watches the cubs, then swims back again and grabs one of the cubs in the scruff, dives with the cub: the little one's first exposure to water. Ten seconds later she surfaces, still holding the cub, heading for a small rocky island, the Coll, about 200 m away. She leaves the cub there, returning almost immediately to her two remaining offspring, which are sitting close together, piping anxiously. Again she grabs one, taking it across to the Coll, occasionally diving with the cub for long periods. But this time she does not come back after landing on the Coll, and the third cub stays behind, whistling loudly, no doubt clearly audible to its mother.

After 20 minutes the cub launches itself into the water, obviously distressed. It swims out, a fluffy ball quite high out of the water, on the smooth surface of the Voe, whistling, and about 50 m out it begins to swim around in circles, apparently totally disoriented. The mother takes no notice; there is no doubt that she can hear the calls of the cub. Slowly, the cub sinks deeper into the water, clearly struggling, still calling, occasionally submerging. Half an hour after it began its swim it finally disappears under the surface.

We made three such observations of cub abandonment in Shetland, all very soon after emergence from the natal holt. It could be, therefore, that such a 'deliberate' reduction of litter size (by neglect) also took place inside the natal holt, before this could be registered by us. We do not know whether particular cubs were more likely to be abandoned than others.

Mortality of Shetland cubs could be assessed from the time they were first seen following the mother around. It was likely that some deaths of cubs in the first few days after emergence went unnoticed, because they may have been out for several days before we spotted a family and could start monitoring them. Of a sample of 33 cubs (18 litters) which we watched from emergence or a few days after, a total of 6 (18 per cent) disappeared and almost certainly died within the following 4 months—mostly, therefore, from causes unknown, though in one case a cub was killed by a farm dog. A further three (10 per cent) died in the next 2 months; after that, when the cubs were about 8 months old, some could

FIG. 8.16 Female otter carrying a cub from a natal holt.

have been large enough to start independent life, and we could not be sure whether disappearance meant mortality or departure. Most cubs became independent from mid-March onwards, at the age of around 9 months, but one only departed when about 13 months of age, in July (Kruuk *et al.* 1991).

Thus, total mortality of cubs outside the natal holt was estimated at 18 per cent for the first 6 months of their life, and from carcass returns (which are probably not very reliable for that age) at 12 per cent for the second 6 months (Kruuk and Conroy 1991).

CONCLUSIONS

In our Shetland study area the average adult female produced 1.02 2-month-old cubs per year, or 102 cubs per 100 females, of which an esti-mated 74 survived up to the age of 1 year . With an (assumed) even sex ratio, there would be 37 females amongst these recruits, which will produce their first litter at 2 years of age. With an estimated female mortality of 8.3 per cent during the second year of life (Kruuk and Conroy 1991), annually 34 2-year-old females would be joining the popu-lation for every 100 adult females present. This compares with a calcu-lated mean mortality amongst females of 31 per cent. Despite the uncomfortably small sample sizes, these data suggest that recruitment and mortality are similar.

On the Scottish mainland the age distribution of otters was compara-ble with that in Shetland, suggesting similar mortality patterns, but litter sizes reported from the British mainland were substantially larger (Mason and Macdonald 1986). Possibly, this difference in litter size is compensated for by a difference in birth interval, which may well be larger in Scotland and England than in Shetland. It was shown by Watt (1993) that young otters along the Scottish west coast are dependent upon their mother for much longer than in Shetland (12 months or more, compared with 8–10 months in Shetland), suggesting longer birth intervals.

Because of the long period of dependence, and hence birth intervals of longer than 12 months, it would be important for mainland otters to be able to reproduce at all times of year (rather than in just one short season). This should enable females to begin the next gestation period soon after separation from the previous litter.

One interesting aspect of the juxtaposition of recruitment and mor-tality in the Shetland otters is the suggestion that both are strongly af-fected by prey availablity. Recruitment is correlated with the summer influx of fish into inshore waters, mortality coincides with the low fish numbers in spring. There is no evidence that these two food variables are in any way linked, but of course they could be under some conditions. Which of the two mechanisms is most important in linking otter populations to numbers of prey must be undecided, as is the present-day role of pollutants therein.

9

Synthesis: some generalizations and speculations

This chapter draws together the observations and conclusions on foraging, habitat requirements, social and spatial organization, and population characteristics. It is suggested that the main problems of otter populations and survival are due to the species' extreme specialization, to an existence in an almost linear habitat, foraging for difficult prey (involving a prolonged learning period), in an energetically very hostile environment which is also highly sensitive to pollution. Otter densities per unit of suitable habitat are comparable with those of other similar-sized carnivores, but numbers in populations are low because of habitat scarcity.

The application of the Resource Dispersion Hypothesis to otter spatial organization is discussed, and the need to consider phylogenetic constraints is emphasized. Fish populations are likely to constrain numbers of otters in several ways; the predators take a large proportion, and food appears to affect both otter mortality and reproduction. Foraging is energetically very expensive for otters because of the high thermal conductivity of the environment, and the need for thermo-insulation in some areas is probably an important constraint on populations. Resource exploitation also affects social behaviour patterns such as scent marking. Suggestions are made for further research on both otter and fish populations.

OTTERS: THEIR REQUIREMENTS AND LIMITATIONS

I will attempt to tie together some of the strings from the various chapters of this book, summarizing results, spelling out implications, and speculating on their biological significance to otters. The results and arguments will centre around habitat, foraging, and population dynamics, referring to the otters' spatial organization and social behaviour. Some points will have been made earlier, but perhaps one should not object to that; all aspects of otter biology are so closely interwoven that it is inevitable that a survey of our present knowledge will lead one repeatedly to the same road intersections.

In these generalizations I will be interested especially in those aspects of otter biology which illustrate more general principles in animal ecology or behaviour, such as the energetics of foraging, or their scent marking, but also I will repeatedly ask the question: how do otters differ from other species? These animals obviously have some unusual problems, their numbers have declined more than most, recovered less than most, and answers to the questions which are posed by these facts will have to be based on comparisons. Clearly the main distinction between otters and other similar-sized carnivores living in the same regions is based on the fact that otters spend much time in water. This relates to their shape, metabolism, locomotion, food, and hence to other fundamental aspects of their biology such as foraging behaviour, social organization, survival, and mortality.

Habitat

At a first glance otter habitat appears to be extremely variable, and the descriptions in this book and elsewhere support this: rivers, lakes, small streams, sea coasts, some with forests along their shores, others without a tree or shrub in sight, in moorland, agriculture, even along dense housing or industry. More detailed studies also suggest that the animals are very catholic in their taste for where they live, swim, hunt, and rear their offspring within each of these scenarios. However, the large variability is deceptive: within this range of different landscapes inhabited by the animals, the actual living space for otters is much more confined, and to some extent all the above attributes are only the rich, non-essential tapestry behind the stage where everything happens.

More accurately, the habitat of otters can be characterized as a narrow strip on either side of the interface between water and land, where food is acquired in the cold, watery inhospitality of one side, and where recovery from this exposure and everything else takes place on the other. The animals' living space differs, therefore, from that of other animals (a) by including both water and land, and (b) by being to some extent one-dimensional, linear. Often we talk about otter ranges in terms of kilometres rather than hectares—even though we know that the width of the strip also counts, and that fishing at depth extracts its energetic toll from otters.

One may consider the otter's body shape as a compromise between adaptations to the requirements of existence on land and in water, with all the resulting limitations. The animal is fairly clumsy on land (compared with, say, foxes, cats, or badgers), and therefore vulnerable to predators. In water it can move fast, but cools rapidly, probably because a really efficient thermo-insulation would interfere with movement on land (as in seals or platypus). To make up for this cooling in water, the energetic requirements of the otter are relatively large and food intake has to be high, which renders otters more vulnerable to prey fluctuations as well as to accumulative effects of pollutants.

Although the overall elongated shape of an otter makes it eminently suitable for a semi-aquatic existence, obviously we should not consider this as an adaptation; it is a common mustelid feature which the otter shares with martens, stoats, and others living miles from water. It is likely that it was this mustelid shape which enabled the Lutrinae to evolve in their aquatic direction, which would have been an impossibility for a canid or felid.

One consequence of range linearity is that animals sharing an area are bound to meet more often than they would in a two-dimensional space. This could affect competition for resources, as I will discuss below; it also affects otters which have a direct interest in keeping away from others. An example is the female with small cubs: she has to avoid infanticidal males, common in many carnivores (Packer and Pusey 1984), and her strategy appears to be to establish a 'natal holt' far away from the water line, from the usual otter habitat. This renders her vulnerable to predation and to more recent threats such as traffic (many roads follow river courses, and they have to be crossed). The special sites in which some natal holts are established are particularly vulnerable to habitat change. Only when the cubs are large enough will families start using shores and banks.

Another important consequence of habitat linearity is that animals have to travel enormous distances: imagine one's own house and garden laid out in a strip of 1 m wide, and the effects of that on distances walked when going about daily business. Much of otters' travel takes place in water, easier because of their morphological adaptations to catching fish, but expensive because of heat loss in this highly conductive medium. It is not surprising, therefore, that any feature of the habitat which enables animals to escape from the consequences of this linearity is highly favoured: shallow coasts (with a wide shelf to catch prey), meanders in rivers, short-cuts across peninsulas and, especially, islands. These are the places where one finds many signs of otters, and where the animals spend a lot of time. To them, any convolution of coasts is advantageous, where there is a long interface with water, which can be accessed by otters from places of safety or with low energy expenditure on land.

When otters live along sea coasts, they themselves are confined to foraging in a narrow strip of water (because of the limitations on depths they can fish), but most of their prey is not. It is a problem which does not arise in most freshwater areas, but in the sea, especially in winter and spring, many of the potential prey of otters move to waters too deep for otters to fish efficiently. Migratory prey for carnivores in land-based ecosystems, for example the ungulates which move through territories of lions and hyaenas in the Serengeti, can to some extent be followed by the predators in a 'commuting' system (Kruuk 1972; Mills 1990; Hofer and East 1993), but this option is not open to otters. In consequence, there is a seasonal low in prey abundance especially along sea coasts, which is associated with high otter mortality. However, there is a bright side to this habitat peculiarity as well: otters are exploiting only the edge of a large resource, and when they have emptied likely hiding places for prey, these will soon be filled up again from the vast store in deeper waters. Thus, prey-catching sites are replenished frequently (probably daily), and this is likely to have a profound effect on foraging strategies (see below).

When a resource patch, for example a tributary with fish, is rich enough to be used by several otters, complications arise over range sharing. In the simplest form this can be mere overlap of neighbouring home ranges, but when there are many such rich patches range sharing becomes difficult. The linearity of the habitat also makes confrontation more inevitable than for animals living in two-dimensional ranges. The otters' options, then, are random dispersion, with tolerance between individuals, or some kind of group territorial system.

For some reason, systems of random dispersion without territoriality are very uncommon amongst carnivores. This contrasts, for instance, with the Australian Dasyurids, carnivorous marsupials in many ways similar to Carnivora, but not territorial (Jarman and Kruuk 1993). It is likely that phylogenetic constraints are playing a role here ('pigs will not fly'); in other words, otters as Carnivora inherited territorial genes, and in their evolution just did not have the option of a spatial system without some kind of territoriality. Of course, there may be distinct advantages to an otter being able to keep a certain set of resources to itself, at least if prior experience of ways of exploiting, of knowing exactly where the fish hide-outs are, is important in the animal's feeding success. Whatever the evolutionary history, otters now appear to have a territorial arrangement in which several individuals share a range, each with its own favourite hunting haunts.

Thus, it is possible that otters' general pattern of a group territorial system, as described in Chapter 2, is not adaptive but merely, or partly, a consequence of phylogenetic inertia, like the overall mustelid body shape. Nevertheless, there appear to be differences in the spatial organization in different areas, and there is variation in group size and range size which cannot be explained by ancestral ties. It is those intraspecific variations in organization which we should seek to explain in terms of adaption to environmental characteristics, and in such explanations the Resource Dispersion Hypothesis (RDH) (Macdonald 1983; Carr and Macdonald 1986) could play an important role.

In simple terms, the RDH explains the existence of group territories as an adaptation to resources, especially food, occurring in relatively small, temporal concentrations or patches. One individual needs several such patches within its range in order to always have prey available, but because patches are rich (though temporal), several other individuals will also be able to feed from them without substantial competition. If there are large risks involved in dispersing from a natal range, then it could be advantageous for an animal to stay in the home group under such conditions, and reproduce there or await future opportunities for dispersal.

This 'RDH explanation' holds fairly well for group-living animals like badgers (Kruuk 1978) or red foxes (Macdonald 1983), but would the RDH have predicted the group organization for otters? We suggested earlier (Kruuk and Moorhouse 1991) that the answer to that is yes, as a vague generality. However, it would not have been possible to predict the details of otter organization from our knowledge of food dispersion, for instance the separate core areas of individuals. Nor would we be able to predict numbers of otters per group range, or even size of ranges, merely from our knowledge of fish dispersion and production. For an appreciation of the extent to which the spatial and social system is adapted to environmental requirements one needs to know many more details such as foraging strategies, resources other than food (for example fresh water along the sea coasts), and, as I argued above, phylogenetic constraints.

The linear nature of the otters' habitat is likely to have important effects on spatial organization. For instance, some animals (for example badgers, Kruuk 1978, 1989*a*) have the option of 'central place foraging'. This means using one site or den as a base and exploiting the home range from there, going off at different times in different directions, foraging in the various patches at suitable times, but always returning again to the central site. This, clearly, is not feasible for otters because of the huge distances involved, except on the relatively rare occasions when they live in a non-linear habitat such as a small lake. They have to move about, basing themselves in different parts of the range at different times, compromising between the need to 'know' the stretch they are fishing, and the need to exploit different areas far apart. The core-area system described in Chapter 2 is one solution to this need for compromise.

Apart from problems of the consequences of habitat linearity, otters also have to cope with difficulties *per se* of spending much time in a wet medium. It is energetically costly because of heat loss, and the otters' land base will have to provide facilities for the animals to recover after swimming: safe holts or couches (for example in reed beds), or, when living in a marine habitat, freshwater washing sites along the coast or inside holts. In many inland areas, dry, underground holts will be an impossibility, but the otters' couch building behaviour makes up for this. In fact, when islands or reed beds are available, otters appear to prefer to rest above ground even in winter, perhaps because their pelts dry better.

There are other, specific risks attached to an aquatic habitat: it may freeze over, making access impossible or very difficult, or dry up. Neither of these risks appear to be a great deterrent to otters; the animals have a range of behaviour patterns to cope with ice, including the use of special entry-holes, or exploiting open patches or running water. Species such as the Cape clawless otter move over large distances when rivers dry up, returning with the rains, when the crabs emerge again (personal observation). Nevertheless, food availability is greatly reduced during such conditions, and it is likely that this constitutes a bottleneck for some populations, although it has not been demonstrated.

Foraging

Unlike most other Carnivora, otters eat mostly fish, but their relationships with prey and prey populations show many characteristics which appear to be general, and applicable to other predator–prey interactions in a more terrestrial environment. However, where otters are unusual is in the cost of their foraging trips, in the huge energetic requirements of keeping warm whilst fishing. All is well if fish occur in sufficient density for otters to be able to get their food quickly and eat, but if they have to search for a long period the expense of foraging increases so rapidly that otter life becomes difficult to sustain. In several areas where otters were studied they appeared to be living uncomfortably close to their energetic limits; a relatively small reduction in prey availability would have made such places unsuitable for otters. This curious situation is caused by a phenomenon which is intuitively difficult to accept, expressed in the 'break-even curve'. The negative relationship between foraging profitability and the time of foraging needed each day to stay alive (when prey is caught in cold water) is not linear, but follows a steep curve (Fig. 6.14, p. 179).

Moreover, whilst often living close to the point where the curve of daily cost of foraging goes up sharply, otters also have to contend with the fact that their staple food is often harder to catch than that of many other carnivores. Compare the skill and energy needed for a terrestrial mammal to catch, for example, a trout, with what a badger needs to get at its earthworms, or a cat to ambush its voles. Hardly surprising, then, that it may take a young otter more than a year to acquire these skills and efficiency, with initially a great deal of what appears to be 'teaching' by its mother. It is highly likely that this long period of dependence affects the breeding interval in some habitats, and therefore reproductive output.

Otters have to eat large quantities compared with other carnivores, and from the many studies on diet it is clear that they do not take just any fish, frog, or crab, but are selective, probably largely in order to balance their energy budget. They do not, or only rarely, chase fast-swimming fish, and they also leave many crabs alone as unrewarding— just concentrating on bottom-living or resting fish species, preferably ones with a high lipid content such as eels. Efficient foraging strategies include hunting at the appropriate time of day (when fish are inactive), in the sea at the right stage of the tide, hunting in shallow water, avoidance of back-tracking, the repeated use of specific feeding patches, harvesting and reharvesting on an almost daily basis with the fish repopulating the empty niches before the otter returns. When watching otters swim, dive, and catch their fish there is no doubt in my mind that they have a detailed knowledge of their beat, just as a fisherman knows the pools and individual rocks where he will cast his fly.

Detailed knowledge of fishing sites and the kind of prey likely to be encountered will probably have important benefits to the otters. It appears that the expectation of success is expressed in the otters' success rates per dive: otters tolerate lower success rates per dive where prey is large or the water very shallow, and only minimal changes in success rate occur between seasons, despite changes in fish availability. In other words, when otters dive they expect a given return for effort; their knowledge of feeding areas enables them to maximize this. It is important, therefore, for animals to be able to use the same resource-rich patches again and without interference from others. This need has implications for the social and spatial structure of the otter population, affecting even behaviour patterns such as scent marking, sprainting.

I have argued that, on the one hand, a Shetland otter needs to use different stretches of a sea coast, because various important prey species do not occur in the same places: some need sheltered areas, some need exposed sites, and different algal communities play a role. But on the other hand, once an otter has a range encompassing such a variety of sites, it can allow others to share, without competition for food, provided that they can avoid exploiting exactly the same patches. This, it appears, is achieved with the organization into core areas, and a signalling system with spraints which informs other otters that someone is already exploiting a site. This enables newcomers to keep out, to the benefit of the animal who was there first as well as to the newcomer itself (the latter because it does not have to waste energy exploring an already part-emptied site).

One may expect, *a priori*, that food is likely to have its main impact on populations of otters and on the behaviour of the animals during periods of shortage or abundance. A first effect to look for is the choice of prey; during shortages, low-preference food will be more in evidence, in the case of otters prey such as rabbits or birds. The role of amphibians, especially frogs, is somewhat ambiguous in this; they are taken mostly, and in large numbers, during times when fish stocks are low, but that is also the time when frogs themselves are much more available to otters. They would appear to be an excellent prey, of the right size and in the right kind of places. It is possible that, at least in some areas, frog populations play an important role at a crucial time of year. Frog populations have declined in many countries (Blaustein and Wake 1990), although this has rarely been documented in detail, and this may be one of the subtle changes in the otters' environment which has made life more difficult for them in the last few decades.

At least one of the staple prey, the eel, shows similar signs of decline, although again insufficiently documented. Eel fisheries in several countries have deteriorated markedly, one of the causes being that smaller numbers of elvers are entering fresh water (Moriarty 1990). This could have affected otters in many areas.

Studies on otter feeding ecology in fresh water show that the quantity of fish consumed by the animals is large in relation to the fish populations themselves. In the rivers of north-east Scotland, where fish populations are dominated by trout and salmon, each year otters eat considerably more than the total 'standing crop' of fish. Of course, fish productivity is high, and larger than the standing crop itself, but even if we express otter predation as a proportion of fish productivity, it is still very large, in fact more than half. This may well be the case also in other areas with good populations of otters, dealing with different species of fish. Our own first results on otters feeding on eels in Scottish lochs suggest that there, too, they take a high proportion of the standing crop. I should add that the predation rates of otters on salmonid fish were calculated for the resident, non-migratory phases of trout and salmon; we did not know what proportion of the large fish, returning from the sea, were taken.

Of course, otters' high rate of predation on fish does not necessarily mean that they actually affect the 'standing crop', the total biomass at any one time. More likely, fish numbers will determine numbers of otters, and we found that there are more otters where there are more fish, a significant positive correlation; fish numbers could be determined by food conditions in the streams. But it may well be that under certain conditions, otters would be competitors for food (fish) with other predators—such as man. This has not been demonstrated anywhere, but the ingredients for it are there.

One complication in the relationship between populations of otters and populations of fish species is the fact that male and female otters vary in their selection of habitat, they live in different waters, and eat different sizes of fish. For example, males are shown to live along more exposed coasts or in larger rivers than females, feeding on larger prey; but there is, of course, much overlap between them. In winter, in those 'male habitats' in mainland Scotland, many large salmon are taken (themselves also mostly male fish). Strikingly, in the 'female areas' in the sheltered bays in Shetland, the main food species, eelpout, showed an enormous increase in numbers in summer, and we demonstrated that this was closely associated with otters' reproductive success. It is possible, therefore, that fluctuations in numbers of given prey species affect male and female otters differently, but many more data are needed before we begin to understand the implications of this.

Populations

Animal ecologists often categorize species, on the basis of their lifetime reproductive strategies, as 'r-selected' or 'k-selected', terms derived from population dynamics terminology. r-selected species produce many offspring within a short time, but they invest little parental care in them, and consequently many of their young die quickly, and the parents themselves tend to be relatively short-lived (for example rabbits). On the other hand *k*-selected species have few young, look after them for a long time, and invest a great deal of energy in that, and they tend to be longer-lived (for example elephants). Otters clearly fall into the *k*-selected end of the spectrum, with an enormous maternal investment in bringing up only a few cubs. Their big problem is that the adults themselves only live for a short time, compared with many other *k*-selected species.

One expects carnivores of the size of otters to have a long life, after surviving the vicissitudes of early independence, and indeed their potential longevity is over 15 years. But mean life expectancy at the age of 1 is only about 3 years, and an average female, even if she breeds every year, will produce no more than two litters, in some areas with on average fewer than two cubs in each litter. There is very little room for manoeuvre with such small numbers, and an only small increase in adult mortality, or a decrease in recruitment of cubs, could seriously affect the viability of a whole population.

The pattern of recruitment and mortality in otters can usefully be compared with that in another well-studied mustelid of comparable size, the badger (Cheeseman *et al.* 1987). There, adult mortality is similarly high, but it does not increase with age as in otters, and most deaths occur in young animals. Females produce more cubs per year and per lifetime than do otters, which is then followed by high cub mortality. This suggests that badger recruitment could be more adjustable to food or to adult population density than the recruitment to otter populations.

In otters both adult mortality and cub survival are affected by fish availability, at least in some populations but probably everywhere. However, these two aspects of otter populations are not influenced by fish numbers in the same way; adult mortality occurs especially during the annual periods of fish shortage, usually at the time when waters are also coldest (so foraging is most costly). Cub survival, at least along Shetland sea coasts, is correlated with fish abundance during the annual glut, in midsummer. So whatever affects fish numbers and productivity, it is likely, along one of these pathways, to have implications higher up the ladder in the aquatic community, in the populations of otters.

Other environmental effects on adult otter mortality, either direct or indirect, by reducing fish populations, may be caused by pollution, which affects aquatic habitats more than most others. It is difficult now to establish which compounds could have been involved in the major decline

of otter populations in previous decades, and some of the possible cul-
prits, such as dieldrin or DDT, have all but disappeared from fish and
otters in Europe. But we still have to account for the slow, or absent, recovery of otters in Britain, and for the fact that otters are still disappearing from many regions in Europe. This is in contrast to many other
species of birds and mammals which were severely affected by pollution,
species such as birds of prey, herons, and bats, which have made a good
recovery.

The effects of pollution have to be considered jointly with other mortality factors such as food shortage, and this has been attempted to
some extent in the Shetland studies. There, a likely scenario was that
otter numbers were limited by food, and that older animals, with high
levels of mercury in their bodies, would suffer the highest mortality. The
role of different pollutants is discussed further in Chapter 10.

A typical characteristic of populations of otters is that the animals
always occur in small numbers—this is true for all species of otter,
except for *Enhydra* which may be seen in 'rafts' of several hundred
animals. Species such as the fox or the badger can attain densities of
up to some 30 per square kilometre here in Britain (0.1–0.3 per
hectare) (Macdonald 1987; Cheeseman *et al.* 1987), but for *Lutra lutra*
we get no more than some 0.8 adults along a kilometre of rich
Shetland coast, or 0.3 per kilometre of stream in a 'good' area in mainland Scotland. Of course, these otter figures also translate into actual
densities of up to 0.5 animals per hectare of fresh water, or 0.1 per
hectare of suitable coastal water: therefore the actual densities per
area of habitat are of the same order of magnitude as those of other
similar sized carnivores. The point is that in any one region there is far
less suitable habitat for otters than for those other, terrestrial carnivores, and otter populations, despite densities which compare well with
other carnivores, cannot fall back on substantial numbers if something
goes wrong.

We still know little of the genetic variability of otter populations, but
several research projects have recently started. For Shetland, with fewer
than 1000 otters, we have argued (on the evidence of throat patches)
that the animals are genetically distinct from the distant mainland
otters, and this could imply that they have derived from a small founder
population. Consequently, there might be little genetic variability, and
therefore greater vulnerability to diseases, as in the case of the African
cheetah (*Acinonyx jubatus*), and other species (Wayne *et al.* 1986).
Fortunately, such threats do not appear to have materialized yet, but
the genetic variability of this and other otter populations clearly needs
studying.

Life at the edge of a precipice

Considering together the threats to otters and their problems as outlined in the previous sections, it is in some ways surprising that otter
populations do still exist. Many of the comments made are valid for
other species as well as for *L. lutra*, and in Fig. 9.1 I have summarized
the main factors influencing otter populations, to put some of the observations in perspective.

We are looking at an animal which lives in a relatively uncommon,
linear habitat, where it has to cover huge distances to satisfy its needs.
It specializes in difficult prey types which take a lot of learning to
acquire them, foraging itself is energetically so expensive that large
quantities of prey have to be caught in a short time, and toxic substances are more likely to accumulate than in other habitats. The

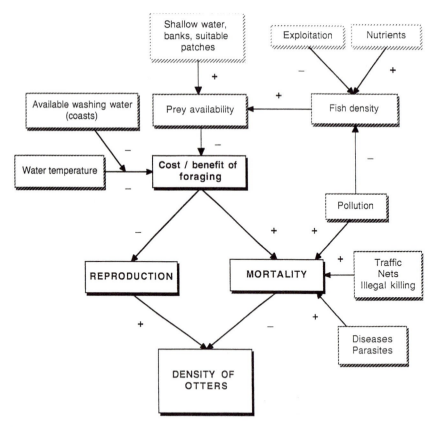

Fig. 9.1 Model of some environmental effects on otter populations. '+' or '−' indicates
whether effects are positive or negative.

underlying relationships are not linear; daily foraging costs in cold water increase almost exponentially with decreasing prey availability, and environmental pollutants such as mercury increase steeply with age. Whether in response to all this or for other reasons, otter populations have a very high mortality and low reproductive rates.

It is not surprising, then, that otter populations are highly vulnerable, and that they have taken some severe knocks in recent decades. Somehow, the animals cope with the dangers, but otters have more odds stacked against them than most other carnivores. Extrapolating from what we know at the moment, it is likely that only a small increase in a pollutant, a decrease in prey population, a severe winter or a drought may have effects much more dramatic than for many other similar-sized species.

To some extent, this vulnerability must be a consequence of feeding specialization *per se*, a vulnerability shared with other carnivores such as the aardwolf (*Proteles cristatus*), cheetah, wild dog (*Lycaon pictus*), giant panda (*Ailuropoda melanoleuca*), and others. However, added to this is the fact that otters have specialized in a habitat and kind of prey that must be particularly difficult for a carnivore, as adaptations to life in two different media have to be compromised.

GENERAL ECOLOGICAL AND BEHAVIOURAL IMPLICATIONS

In a single-species study such as this, there is the clear temptation to look only at how this one focus of our attention is affected by its environment. But clearly, the value of the various research projects on otters, in Britain and elsewhere, stretches beyond the implications for otters only, and for otter management, and I will mention a few examples.

For instance, the relatively very high costs of the hunt, of foraging, are a phenomenon likely to be shared also by carnivores such as wild dogs, with their long, very fast chases (Estes and Goddard 1967; Skinner and Smithers 1990), and the American badger *Taxidea taxus*, which digs large and deep holes to obtain its large rodent prey (Lampe 1978). In such species we are likely to see a model of foraging energetics comparable in shape to that for otters (p. 179), albeit with different parameters. In such models, factors extending the length of time spent hunting for every unit of energy gained can have inordinate effects on overall profitability of hunting; these could be factors such as vegetation characteristics for foraging wild dogs, or the ease of soil excavation for digging American badgers.

Scent marking of resources is very common amongst carnivores and many other mammals; frequently, it is associated with territorial behaviour, the olfactory message of 'keep out' backed up by aggressive behaviour (Gosling 1982, 1990). The sprainting behaviour of otters, however, suggests that no such aggressive back-up is necessary for this type of communication system to operate effectively, and this may also apply to many other scent marking behaviours in other, different species.

The absence of territorial behaviour in Dasyurids such as the Tasmanian devil and the various quolls, animals otherwise so similar to Carnivora (where territoriality is characteristic), is very striking. It is one of the pieces of evidence which suggests that territoriality, this interface between social behaviour and habitat, is at least to a large extent determined by an animal's phylogeny, and not just by environmental requirements (Jarman and Kruuk 1993). If this is the case, then it is not surprising that the very general Resource Dispersion Hypothesis (Macdonald 1983) has rather little predictive value for a species such as the otter. It could mean that the complicated social system of otters, of separate male and female areas, larger male ranges, female group territories with individually separate core areas, is at least to some extent an adaptation to requirements in evolutionary early days.

Otters may make the best of a bad job in this, showing intraspecific variations in the spatial arrangement which is adapted to local requirements. But it could well be that if, for instance, the basic spatial organization of these animals had been random dispersal instead of territoriality, with complete tolerance of other individuals, using a scent marking system as above, the areas would have supported similar or larger numbers of otters. Of course, this is only speculation, but when other species can get by without costly territorial behaviour, such speculation is not total fantasy.

The Resource Dispersion Hypothesis is helpful in understanding spatial organizations, especially the occurrence of social groups, and when predicting the response of a species' spatial arrangements in other, different environments. The otter studies suggest, however, that it is difficult if not impossible, to use only environmental parameters (for example patchiness of resources) to predict the spatial organization of an unstudied species. In a model of factors underlying such organization, the phylogeny, the details of hunting behaviour and other species-specific characteristics have to play an important role.

FURTHER RESEARCH

Several important questions need to be addressed, to cover important conservation problems or because of more academic interest, but usually for both reasons. For instance, at the more extensive end of the scale, we still know little about actual otter numbers over larger areas, about population sizes, and about changes in the areas where previous estimates have been made. Following on from that, the genetic composition of populations needs studying, that is differences between otters from various regions. We need to know what are the barriers between regions, and whether there are differences which should be kept in mind when considering transplantations. What is the genetic diversity within populations: are there dangers associated with inbreeding?.

The processes of population change still need further elucidation. For instance, what is the cause of the increase in mortality with age, how is the oestrus cycle synchronized with the seasons in some regions, and apparently random elsewhere? What is the profitability of foraging in various areas, taking into account success rates and prey types, as well as water temperatures? Much more information is required on populations of relevant fish species, their population dynamics, fluctuations, and responses to predation. The characteristics of fish populations will be crucial to otter populations everywhere, and a dedicated, interdisciplinary approach to the role of fish populations is much needed.

Finally, much more information is required on intra- as well as interspecific variation in the spatial organization of otters, and underlying environmental differences. For instance, how is *L. lutra* organized in areas in south-east Asia, where it lives in totally different vegetation types, where it has to compete with other species of otters, and where it may be subject to much higher predation? Such questions of intraspecific differences need to be answered in order to establish the flexibility of a species in the face of environmental change. Similar questions should be asked for other species of otters—as just one example, what causes the spotted-necked otter to live in large, diurnally active packs in Lake Victoria, and as nocturnal, solitary individuals in streams of South Africa?

Then there are the direct conservation problems: how do the animals respond to transplantation (reintroductions), how do populations respond to increases or decreases in their food supply, what are the pollutant levels in otters from stable, or increasing or decreasing populations?

In fact, the most basic questions on the ecology and behaviour of most species of otter (in areas other than Europe and North America) are still totally open, and a great deal of fascinating research has yet to be done to serve as a solid basis for conservation management.

10

Otters, man, and conservation

Otters are protected by law in many countries, but some persecution still continues, and a few types of traps are described. The motivation for persecution is that the animals may cause damage to fishing gear, and they may take a large proportion of the production of commercially valuable fish populations. The skins of several otter species are still being traded. Nevertheless, persecution is unlikely to threaten otter populations in Europe, but there is more concern for tropical species. The threats to otters as perceived by observers in countries world-wide are quantified; the most commonly mentioned hazard is habitat destruction, with pollution recorded as a second cause in Europe, and persecution elsewhere. Declining prey populations are a likely major hazard, and conservation management should aim at preserving whole ecosystems, focusing on fish populations, in which otters as top predators play an important and high-profile role. Pollution is likely to be an important determinant of survival of otter populations, in conjunction with food availability, but there is still much uncertainty about which compounds are most important; the case against mercury and PCBs is discussed. Important habitat components, as the foci for conservation management, are freshwater pools and streams along coasts, and islands, reed beds, and bank vegetation inland. Overall, the need for an ecosystem approach is emphasized, rather than a single-species otter conservation policy.

PERSECUTION AND EXPLOITATION

Recent results from studies of otter ecology and behaviour will be relevant in the decisions and practice of conservation management. A great deal has been written already about otter conservation, with good summaries in Chanin (1985) and Mason and Macdonald (1986); however, to put the ecological research into a conservation perspective, I will briefly discuss some aspects of past and present relations between man and otters.

Otters, especially in Europe, have achieved remarkable popularity recently, and although few people see them, they must be the most appealing species of mammal in many countries. However, in the not too distant past, say more than 50 years ago, they were considered as vermin, no better than foxes or polecats. In ecological terms, this perception is perhaps best interpreted as direct interspecific competition between predators: otters preyed on the same species in which fishermen were interested, and interfered with fishing nets.

Now, in a more tolerant age and with few otters left, conservationists tend to deny what were seen as the otters' evil deeds, but perhaps it should be acknowledged that the view of the countrymen of old was not totally off the mark, at least not everywhere. In our own recent research in Scotland we recorded otters eating many salmon, often large ones, and they consumed a large proportion of the productivity of fish populations, as documented in Chapter 4. This did not necessarily cause competition with man, of course.

Fishermen in Europe have suffered damage to nets, caused by otters, and I have also seen this elsewhere. In Thailand, for instance, where we used gill-nets to catch fish in rivers, otters (almost certainly *L. lutra*) frequently tore up the nets (they were actually seen doing so), and they took much of the catch, as did *Lutra maculicollis* with gill-nets during my work on Lake Victoria. In Rwanda, this last species took an estimated one-seventh of the catch from the nets of local fishermen (Lejeune 1989). Little wonder that people whose livelihood depends on fish catches show less sympathy for the animals than do the urban viewers of spectacular TV documentaries; persecution of otters is based not just on prejudice. Of course, this does not imply that I approve of persecution in any way, nor do I think that it is effective in preventing damage.

Otters also lower their popularity ratings by taking poultry—ducks, hens, and geese. My own three ducks, free-roaming in the garden in Scotland, were taken from the pond one snowy March, by an otter female or small male. They disappeared one at a time at night, over a couple of weeks, carried off and eaten in the marsh about 1 km away, with good tracking snow to record it all. A well-known large male otter regularly raided the hen-house of one of my PhD students on Skye, Paul Yoxon, who is studying the animals along the coasts there. Again on the Scottish mainland one of our radio-tagged otters, followed by Leon

Durbin, regularly foraged from a collection of captive ducks (whilst the owner blamed mink). Shetland farmers frequently lose hens and ducks to otters, and I recorded many other incidents, including two geese of a nearby game-keeper here on mainland Scotland.

Similarly, fish-farms may occasionally be visited by otters, if there are chinks in the armour of the ponds or floating fish-cages (Fig. 10.1): holes in the protective netting, or absence of antipredator wiring. Then, otters may take substantial numbers of fish, although usually fish-farms are easy to protect, and ignored by the animals. In Shetland, where about half-way through our study a large, well-protected fish-farm was established right in the centre of our intensive study area, we frequently saw otters swim closely past the floating cages full of salmon, never taking any overt interest in them. However, in South Africa (southern Cape province) I watched a Cape clawless otter for about half an hour attempting, unsuccessfully, to get into a floating fish-cage similar to the Shetland ones: walking on the board-ways and biting at the netting.

In the face of such evidence, conservationists do not help their cause by denying the 'crimes'. Otters do cause problems for people living in the countryside, and that is the reason for the animals' persecution, now fortunately rare and illegal.

FIG. 10.1 Floating fish cage, in a Shetland salmon farm at Lunna. Potentially vulnerable to otter predation, but proper application of guard nets is very effective in practice.

The fur trade also provided a strong incentive to kill otters. Pelts are highly valued, and trapping river otters *L. canadensis* is still an important source of income in some areas of Canada and the United States (Foster-Turley *et al.* 1990). This species escaped the fate of the sea otter *Enhydra lutris*, which was hunted almost to extinction for its fur. The sea otter is much easier to catch than the other species, and it also has a more valuable, thicker fur; if it had not been for its legal protection in 1911 there is little doubt that it would have disappeared entirely. It is now once again common along the Pacific seaboard of the US, Canada, and Russia (see Riedman and Estes (1990) for the history of its persecution and recovery). In tropical countries the present-day trade in otter skins is likely to have serious effects on otter species, for example on the giant otter in South America (Duplaix 1980; Foster-Turley *et al.* 1990).

In Europe the killing of otters for fur has virtually ceased, with legal protection for the species in most countries (Foster-Turley *et al.* 1990), and with strong social pressures from conservationists. But even now skins are still popular for decoration in the house, and in Scotland the 'mask' is a popular cover for the traditional sporran. In markets in southeast Asia many skins of *L. lutra* are still being sold, legally or otherwise.

The most impressive historic monuments to the role of otters in the fur trade in Europe are in Shetland. Scores of unique and ingenious 'otter houses' were constructed by the Shetlanders, and can still be found there, more or less intact: well-built, stone structures at strategic places just above the high tide mark (Fig. 10.2). They were made to catch otters by the construction of what is almost a small artificial cave, the entrance with a trap-door, a sprainting stone or couch site inside, a 'lid' in the ceiling to kill and remove the animal. I talked to several old Shetlanders who caught otters with them. Many of the structures must be centuries old, abandoned by the people who used them, but still popular with the otters of today, who sleep and spraint inside the otter houses as they have done over the ages.

Traditionally the most common method of catching otters on mainland Scotland was the leg-hold or 'gin' trap, with its sharp teeth, now outlawed in Britain and many other countries but still in common use in North America. I still find the odd one, illegally employed here in Scotland and set for foxes. Clearly, otters would be highly susceptible to being caught by this type of trap, or by neck snares, set on one of their well-worn runs near a spraint site. Probably the nastiest otter trap I met was in Norway, where a fisherman showed me an underwater trap, spring-loaded, with huge spikes to go through the body of the animal, set off by the otter on touching a piece of string when diving close to a holt or spraint site (Fig. 10.3).

FIG. 10.2 Remains of a traditional Shetland 'otter house', a permanent and highly effective trap to catch otters.

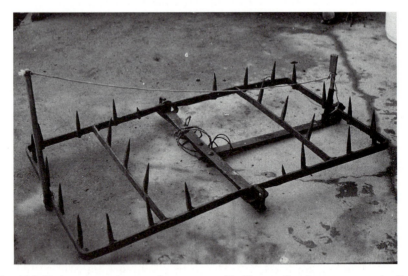

FIG. 10.3 Underwater spring trap for otters, used in Norway. Set close to an otter landing site, with the two large jaws activated when an animal touches the string.

Shooting otters, after flushing them with dogs from their couch or lying-up place, also accounted for considerable numbers of otters killed here in Scotland by gamekeepers. For instance, along a 20 km coast of a peninsula in north-west Scotland, one keeper shot 93 otters in less than

15 years (Kruuk and Hewson 1978). This cull stopped 3 years before we began to study the animals there, when the population appeared to be thriving. But otters are still shot along many Scottish rivers, despite full legal protection (Fig. 10.4). Not surprisingly, we only occasionally come across the odd victim, because the evidence is so easy to dispose of. Some years ago a beaver escaped from a wildlife park in north-central Scotland, and was shot by a keeper who thought it was an otter.

Hunting of otters for sport, with hounds, must have been motivated originally by arguments of vermin control, but was outlawed in Britain in 1981. The successes of the 'Otter Hunts' in killing otters were well-documented, and the hunts between them throughout England accounted for some 200 animals each year (Chanin and Jefferies 1978; Jefferies 1989). These authors argued that numbers killed by hunts were unlikely to have made serious inroads into otter populations at the time (although it may have altered the age structure) (Jefferies 1989). This is probably also true for most of the other methods of vermin control employed against the species in Europe.

Nevertheless, mortality caused directly by man could be significant when added to the more recent, less immediate causes of population decline. Killing otters had to be made illegal, even if for that reason alone. But although it is now outlawed in many countries, otter conservationists in Britain and world-wide should remember that there were,

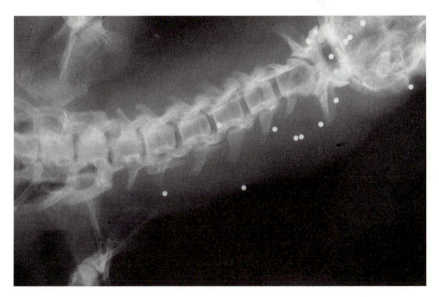

FIG. 10.4 X-ray photograph, showing shot-gun pellets in an otter killed along the River Dee, north-east Scotland 1991.

and are, strong motivations behind the persecution of otters, and these ideas are not likely to disappear quickly.

THREATS TO OTTER SURVIVAL

Increases in particular single causes of mortality are not necessarily threats to populations, even when such populations are at low density, just 'holding their own'. There are many compensatory mechanisms operating in all animal populations; for instance, if a population of some species of mammal is limited by food, a quite dramatic increase in traffic mortality may still have no effect on densities. This should be kept in mind when reviewing the many reasons which are mentioned as causes of decline.

Recently, Foster-Turley *et al.* (1990) outlined the threats to otter populations in countries all over the world, as perceived by a large number of observers with local knowledge. I analysed this very useful set of descriptions, by continent; the results are given in Table 10.1. Overwhelmingly, habitat change is seen everywhere as a major cause for concern; in Europe the next most important one is pollution (Fig. 10.5), but elsewhere it is more often direct persecution.

TABLE 10.1 The threats to otters in 102 countries. Data from the IUCN report by Foster-Turley *et al.* (1990), as per cent of countries in which a given environmental change was perceived as a threat to future survival of otter populations. All species of otters combined.

	N	Hab	Poll	Acc	Pers	Fish	Recr	Hum
Europe	29	**90**	**59**	45	41	31	34	3
Africa	34	**71**	15	0	**38**	18	3	24
Asia	19	**89**	42	11	**53**	11	0	5
Americas	20	**75**	35	15	**40**	10	0	0
Total	102	**80**	36	18	**42**	19	11	10

N = number of countries from which information was received. **Hab** = Habitat change, **Poll** = pollution (direct effects on otters), **Acc** = Accidental deaths (traffic, drowning in fish nets or traps, etc.), **Pers** = direct persecution (shooting, trapping), **Fish** = decline of fish populations (over-fishing, pollution), **Recr** = disturbance caused by recreation and tourism, **Hum** = direct disturbance from people (other than recreation).For each continent the two most frequently quoted threats are given in bold.

FIG. 10.5 The potential for disaster: wreck of the giant oil tanker *Braer* off the south-east coast of Shetland.

Obviously, these data have to be viewed cautiously, as they are not based on accurate observations but on subjective perceptions, often coloured by expectation. On the other hand, however, the experience of many good naturalists has gone into this table, and the figures are likely to be a good indication of the truth.

Absent from the list of expected hazards is disease, about which we know little for otters. The animals are susceptible to several diseases which also occur in pets and livestock, and it is therefore possible that increased contact with man will lead to an increase of such diseases as distemper. There were fears that a recent outbreak of phocine distemper in seals would affect otters in marine habitats around Britain, but this did not materialize (Kruuk *et al.* 1989). In our study area in northeast Scotland an otter aborted, caused by a bacterial infection which also occurs in sheep (Weber and Roberts 1990). In general, however, there is no good evidence that disease plays a substantial role in otter population dynamics.

In the responses to the survey, the details given about what I have generalized as 'the threat of loss of natural habitat' included damming up and canalization of rivers, removal of bankside vegetation, construction of dams and barrages, and many other onslaughts on nature. Looking at this from a cautious angle first, we do not know what the effects are of such changes on otters; observers feel intuitively that a river which is straightened out, or has its trees removed from the banks, is a less favourable habitat for otters, and that a barrage or a dam, ugly as it is, must be unsuitable for these animals. Moreover, otters tend to spraint near trees, on conspicuous rocks or logs, so one finds fewer signs of otters if these attractive parts of the landscape are removed—but it does not necessarily mean that there are fewer animals.

An example of the need for caution. The establishment of floating fish-farms along the coasts of Scotland is specifically mentioned as a threat to otters in the IUCN document. The fish-cages are ugly, often spoiling beautiful wild coasts—but in our intensively studied areas on Shetland we found no effect on otters (comparing before and after the arrival of the fish-farms), neither on numbers nor on movements or activity patterns. In general, we need much more research on the actual impact of changes, on densities of otters before and after the many habitat changes about which we are concerned.

However, there can be little doubt that otter habitat is extremely and unusually vulnerable to man-made changes, and that at least some of those changes could seriously affect numbers of animals. The extra vulnerability of the habitat is due to (a) its linearity, with otters needing very large ranges, (b) the importance of water, which is also badly needed by man, and (c) the importance of bank vegetation, in areas also sought after by man.

In Britain, the disappearance of reed beds is one of the changes most relevant to otters; there are very few left of any size (Everett 1990), but where they still occur, otters use them a great deal for resting, breeding, and feeding (Chapter 2). Along coasts, the most serious habitat threat is the drainage of freshwater pools and small streams, which otters need to wash themselves (Chapter 7; Beja 1992). Without this fresh water, otter life is virtually impossible in coastal areas.

Nevertheless, habitat change is unlikely to have caused the crash of otter populations in Europe. The vegetation and appearance of many areas, such as the lakes in southern Sweden where Erlinge carried out his otter research in the 1960s (Erlinge 1972), or the lakes in Holland, the streams in southern England, are all virtually unchanged, but the otters are gone. Even if there were changes, we know from radio-tracking work (Green et al. 1984; Durbin 1993) that otters are highly catholic in their choice of streams and banks, and are unlikely to be affected by removal of trees, straightening of banks, or by agricultural activities.

In fact, although otters are often seen as an 'indicator species', a symbol of clean water and natural habitat, they are remarkably tolerant of a great deal of anthropogenic mess, such as buildings, rubbish in streams, disturbance, and noise. Our radio-tagged otters here in Scotland frequently slept in or under man-made structures: rubble, concrete blocks on banks, even in an old car in the stream close to a farm. In Shetland, otters slept right inside the pump houses at the huge oil terminal at Sullom Voe. Seeing otters playing and splashing about in the waves, it is clear that their image may be useful as that of a symbol, of an ambassador for a clean environment, but a good indicator species they are not: they are too tolerant.

Fish populations are part of the otters' environment, and we have shown that otter numbers are closely correlated with fish biomass. For non-specialist observers it could be far from easy to assess a decline in fish populations, let alone see it as a possible reason for the decline of otters. Nevertheless, world-wide it was seen as a cause for concern for otters by almost one-fifth of the respondents in Foster-Turley *et al.* (1990). This concern may well have been an under-valuation, given the decline in eels in Europe (Moriarty 1990), of salmon (Jenkins and Shearer 1986), both major food species, and of amphibians (Blaustein and Wake 1990), which may be vital to otters at times of food stress in spring (Weber 1990). The introduction of the American red swamp crayfish (*Procambarus clarki*) into Europe, especially into Spain and Portugal, has been reported to be of great benefit to otters (Delibes and Adrian 1987). In general, however, and also from our intensive studies, it appears that the disappearance of suitable fish and other food species is a major threat to otters.

In Europe pollution is the most frequently mentioned serious hazard to otter populations after habitat deterioration, but in countries elsewhere in the world there appeared to be less concern about this (Table 10.1). There is little doubt that pollution caused the huge population crash in Europe, in otters (Jefferies 1989) as in birds of prey, herons, grebes, kingfishers, and others (Newton *et al.* 1993). Several important questions remain, however, especially: (a) which compound(s) were responsible, (b) is the lack of otter recovery due to the same cause, and (c) does pollution affect otters directly, or work mostly through its effects on prey populations?

At times the role of pollutants in the life and death of otters is thrown into doubt, when one sees the animals swimming and surviving in rather dirty waters. Around Shetland's oil terminal (Fig. 10.6), with one of the densest otter populations, the animals sometimes swim through a thin, oily sheen on the sea. Many rivers and coasts in Europe, South Africa,

FIG. 10.6 Prime otter habitat in Shetland: the Sullom Voe oil terminal.

and elsewhere I have seen to be horribly dirty and with much debris, almost bound to be chemically polluted in many ways, but along the banks there are otters' spraints. During our study even in Shetland domestic waste was tipped directly into the sea (Fig. 10.7), but otters were around in the immediate environment. Perhaps one should see these as 'incidents', which can be individually explained, and which cannot detract from the general picture of pollution as a major cause for population decline. But these observations do leave open the possibility that otters are able to tolerate considerable contamination before this affects populations, and this is supported by our data on pollutants in otters in thriving populations in Scotland (p. 220).

There has been a lively debate about exactly which compounds in the environment pose the greatest danger to otters, and there is no clear answer at present. The main contenders are the organochlorines dieldrin (HEOD), polychlorinated biphenyls (PCBs), and a heavy metal, mercury. Of those, dieldrin is generally held to be responsible for the big crash of British populations of otters (Jefferies *et al.* 1974) and many birds (Newton *et al.* 1993), but it has now largely disappeared from the environment. Mason (1989) and Mason and Macdonald (1993) argue that PCBs are currently the dominant environmental hazard, and their considerable experience with aquatic pollutants lends much weight to the discussion. PCBs enter ecosystems from the air or directly in water; their manufacture has officially discontinued now, but they are still present in the environment, occurring in sewage, in industrial wastes, and as products of incineration of some plastics.

FIG. 10.7 One cause of marine pollution: domestic waste tipped directly into the sea.
Unst, Shetland, 1988.

As Mason (1989) points out, PCB levels are generally higher in otters
from countries where the species is declining, in countries with ad-
vanced industrialization and intensive agriculture, such as Holland,
Germany, and many other regions. However, these areas are also likely
to have many other disadvantages for otters, such as high levels of other
pollutants, high traffic densities, and less suitable habitat. Furthermore,
in our studies individual otters have been found in excellent condition
and reproducing with high burdens of PCBs. Some thriving populations
of otters (for example ours in Shetland) have levels of PCBs much
higher than those from areas mentioned by Mason (1989) where otters
are declining or have disappeared, and much higher than the levels
causing reproductive failure in mink (Jensen *et al.* 1977). Finally, PCBs
do not appear to accumulate with age in otters (p. 222), which suggests
that the animals are able to excrete or metabolize these compounds.

This does not mean that PCBs are harmless to otters, but the case
against them is not overwhelming. Mercury may be much more important to the animals than has been appreciated so far; it is highly poison-
ous to man and beast, and unlike the PCBs it accumulates in wild otters

with age (Kruuk and Conroy 1991), increasing to levels of about half the concentrations which are known to kill otters in acute doses (O'Connor and Nielson 1981).

Mercury occurs naturally in many waters, especially in the sea, but it is also an aerial pollutant generated by heavy industry (Lindquist 1985), it is associated with sewage, and it was used as a seed-dressing in agriculture. Many sea mammals can deactivate mercury in their bodies, in a mechanism involving selenium (Koeman *et al.* 1975), but this mechanism does not appear to be present in otters (Kruuk and Conroy 1991). It is possible, therefore, that otter populations are highly sensitive to mercury. In Britain, levels of mercury in grey herons have declined sharply since 1970 (Newton *et al.* 1993), and these birds have made a good come-back since their earlier decline. As herons have many prey species in common with otters, something similar may have happened in otters as well, and perhaps they are now also able to repopulate. Mercury levels in Scottish otters are higher in areas where populations are at risk (close to regions where the animals declined), or at least have been at risk in the recent past (for example in the south). However, they are lower where the animals are plentiful and safe, such as in Shetland and in Grampian, north-east Scotland. This contrasts with PCBs, which are highest in the far north, Shetland. Unfortunately, our knowledge of the distribution of these pollutants in otters throughout Europe is still only fragmentary.

Thus, although most observers are agreed that pollution is a major hazard to populations, the details are far from clear, and some otter populations survive in the face of considerable contamination by compounds which they can metabolize.

From our mortality data we concluded that food availability plays a major role in otter survival and reproduction, and is likely to interact with environmental contamination in its effect on populations. In other words, individuals may be able to cope with levels of several kinds of pollution, as long as food is plentiful. However, if contaminants also affect numbers of potential prey, it could then result in a two-pronged assault on the predator: this is still uncharted territory. For the present purpose, the important observation is that at least some, maybe all, otter populations appear to be food-limited, despite considerable pollutant burdens in individuals.

CONSERVATION OF OTTERS

The general message for conservation management, which should come through loud and clear from the recent research on otters, is that to maintain populations animals will have to be protected in their context, as part of an ecosystem, as part of a food web; we have to think in terms of whole wetlands and river catchments. Nevertheless, public interest tends to focus on species such as the otter, rather than on for example the fish they may eat, and this needs to be acknowledged. I will therefore outline some of the implications of the studies discussed in this book for the conservation management of otter populations, treating otters as 'ambassadors' for wetland conservation. Under this flag of 'otter conservation' we are addressing wider issues of wetland and river catchment management.

On an immediate, practical level there are decisions to be taken about reintroduction of otters in areas where they were formerly abundant, but are now scarce or absent. Should one release captive-bred otters, or animals captured elsewhere, in areas such as Holland, Switzerland, or England? Yalden (1993) reviews the pros and cons of reintroduction for different carnivores, and emphasizes the need for environmental preparation, for suitable conditions to be met before release, for sufficient numbers of animals to be introduced simultaneously, and other prerequisites. Once the environmental problems which caused the otters' local extinction have been overcome, the release of captives could well speed up the return of the otters in isolated areas. However, each introduction should be approached with great caution, as emphasized by the failed reintroduction of sea otters (Estes *et al.* 1993). The argument emphasizes again the need for an approach to the problems of a complete ecosystem, and reintroductions can play a part in this type of management. However, on their own they should not be seen as a quick solution to an extinction problem.

Obviously, all such manipulations of animals will have to be closely monitored by radio-tracking of the releases. The first time this was done adequately for *Lutra lutra* (in 1991–3 in Sweden, by Thomas Sjöåsen) (personal communication) released otters were found to suffer high mortality (mostly from traffic and probably starvation), and especially the captive-reared ones—translocated wild otters fared somewhat better.

If it is decided to release otters in an area, it is important that animals are selected so as to maintain genetic differences between populations. Thus, the origin of the animals to be released should be reasonably close, in terms of otter dispersal; for instance, otters from Ireland or Shetland should not be used to repopulate areas on the British mainland, but it is probably safe to assume for instance that when otters were still common, Czech populations were genetically fairly close to those from Switzerland and Holland, or Norwegian ones to those of Sweden.

One practical implication of the observations on otter food, foraging, populations, and pollutants, is that if we want to manage an area for these animals, we need to ensure that there are sufficient populations of prey species. Again and again, throughout our observations in Shetland, and in rivers and lakes in mainland Scotland, or on other species of otters in Africa and Asia, we found strong suggestions that the animals are food-limited. Facing declining fish populations, waters could be made more suitable for otters by improving the food supply: for instance by establishing and stocking fish-ponds (as suggested for Austria by Foster-Turley *et al.* (1990)), or by improving the trophic status of streams (for example by liming the more acid streams, or even by adding some fertilizer). Anything which adversely affects fish populations should be discouraged: intensive fishing, and especially pollution. Where otters are dependent on specific aspects of the habitat for their fishing (for example rough banks and riffles in rivers (Carss *et al.* 1990), or semi-aquatic scrub for the spotted-necked otter (Kruuk and Goudswaard 1990)), the management of such parameters of food availablity would be important. Obviously, the effects of such methods would have to be closely monitored.

Clearly, some of the methods which could be used to 'promote' otters would be highly artificial; maintaining specially stocked fish ponds for otters amounts to otter-farming. Also, there could be important side-effects of 'improvements' for otters to other aspects of the environment. For instance, a species such as the Cape clawless otter may benefit from eutrophication of streams, as this would encourage crabs, and it may be that the introduction of American crayfish into Europe has boosted otter numbers—but the wider effects of such changes on vegetation on other species of animals could be serious.

Other results from recent research on habitat usage by otters are relevant to otter conservation. Trees along river banks may be attractive to us, but appear to have little effect on otter dispersion or movements; on the other hand, islands in rivers or lakes, even artificial ones, attract the animals from a long distance, so may well be important. Favourite vegetation for resting and sleeping includes thick bushes such as bramble, gorse, and in Scotland often rhododendron. If one is going to design 'otter havens' these habitat preferences could be highly relevant. A bank of thick brambles may attract animals more than a carefully designed 'artificial holt', because many otters sleep mostly above ground. Conclusions on habitat preferences of otters have often been based on spraint distribution (for example Jenkins 1981), but we now know that this may lead to false deductions.

An observation on habitat use which recurs frequently, for the Eurasian otter and several other species, is that they do not appear to avoid houses, industry, roads, camp-sites, and much of the mess left by people (Figs. 10.6, 10.7, 10.8). Purely in the name of otter conservation, it is difficult to argue that such riparian development is detrimental—usually, otters will tolerate it.

A specific habitat issue has been identified along sea coasts: otters cannot do without small streams and freshwater pools, or holts going down to groundwater level. They need frequent baths to be able to survive the consequences of sea-water foraging. Development such as drainage which reduces access to fresh water, is likely to have an immediate effect on otter numbers, and in areas such as Shetland, north-west Scotland, Ireland, and Portugal it is important that coasts which are well-provided with such pools and/or peaty soils are adequately protected. At present, most of the really 'good' otter coasts in any of those areas, and to me most obviously in Shetland, have no protection of any sort. Conversely, our results suggest that otters in other areas (for instance Orkney) would benefit from artificial freshwater pools along the coast, which should effectively increase numbers.

All these points do not detract from the fact that what is required, most of all, is a conservation policy for whole wetlands. The above comments are merely suggestions to pursue the restricted aim of

FIG. 10.8 No qualms about human proximity: young male otter on the harbour wall, Lunna, Shetland.

maximizing numbers of otters—probably, we now have a substantial proportion of the knowledge required to follow such a course. But an example, albeit a rather exotic one, demonstrates that merely maximizing otter numbers is not necessarily a desirable management option.

One of the most natural, best-managed areas of wilderness in Africa must be the Kruger National Park, in the north-east of South Africa. Through it, the Sabi River runs its course, a permanent river with sand banks and rocks, beautiful pools, thick scrub, reeds or grasslands on the banks, and masses of fish, as well as crabs. Otters, however, are few; they are seen only rarely, one finds the odd track of a Cape clawless otter on the many sandy places, but no sign of the spotted-necked. There are many places in southern Africa where otters are much more abundant, in rivers exposed to agriculture and human disturbance. The best explanation for the difference is the presence of many crocodiles in the Sabi river, and their virtual absence wherever rivers are not protected. Exactly the same case, the same observations, can be made for the Mara river, in the Serengeti National Park and Mara Game Reserve in East Africa.

Crocodiles also need careful conservation management and protection, and so do many other species which share their habitat with otters. Clearly one should not consider the conservation of otters in isolation from the rest of the fauna: an area with many otters is not necessarily the conservation goal we should be aiming for. The same points can be made about areas elsewhere; in Europe, for instance, acid and nutrient-poor bogs are unsuitable otter habitats, as they have very few fish. However, it should be possible, in theory, to attract the animals there, given some agricultural eutrophication, which would bring invertebrates, and fish. Once there are eels, otters could follow—but we would have lost a whole, fascinating oligotrophic community. In order to manage an area, a wetland or a river catchment, or a coast, one has to decide first what kind of community is desired; otters are not necessarily part of that in every case.

Because of the size of areas used by top predators such as the ones I am discussing here (up to 80 km of stream for one individual otter), a strong human influence, including agriculture or fishing, will almost necessarily have to be included in any management plan. It is possible, however, to accommodate this next to an impressive diversity of wild fauna, and I believe that it is one of our more important duties as research scientists to advise on how this can be done.

Questions need to be addressed such as how many fish one can harvest before affecting numbers of top predators, how nutrient input from agriculture and forestry affects fish populations (through plankton and invertebrates), how organochlorines, mercury, and other pollutants affect the food web. One needs to know much more about these problems and several others before we can feel some confidence that we are managing rationally. I hope that at least some of the conservation agencies in Europe will direct funding towards these ends, because rational management of the European wetlands is vitally important.

References

Anderson, S. S. (1981). Seals in Shetland waters. *Proceedings of the Royal Society of Edinburgh (B)*, **80**, 181–8.

Andrews, E. and Crawford, A. K. (1986). *Otter survey of Wales 1984-1985*. Vincent Wildlife Trust, London.

Arden-Clarke, C. H. G. (1986). Population density, home range and spatial organization of the Cape clawless otter, *Aonyx capensis*, in a marine habitat. *Journal of Zoology, London*, **209**, 201–11.

Bacon, P. J., Ball, F. G., and Blackwell, P. G. (1991). A model for territory and group formation in a heterogenous habitat. *Journal of Theoretical Biology*, **148**, 445–68.

Baker, J. R., Jones, A. M., Jones, T. P., and Watson, H. C. (1981). Otter *Lutra lutra* L. mortality and marine oil pollution. *Biological Conservation*, **20**, 311–21.

Bartholomew, G. A. (1977). Energy metabolism. In *Animal physiology: principles and adaptations* (ed. M. S. Gordon), pp. 57–110. Macmillan, New York.

Beach, F. A. and Gilmore, R. N. (1949). Response of male dogs to urine from females in heat. *Journal of Mammalogy*, **30**, 391–2.

Beja, P. R. (1992). Effects of freshwater availability on the summer distribution of otters *Lutra lutra* on the southwest coast of Portugal. *Ecography*, **15**, 273–8.

Bekoff, M. (1989). Behavioural development of terrestrial carnivores. In *Carnivore behavior, ecology, and evolution* (ed. J. L. Gittleman), pp. 89–124. Comstock Publishing Associates, Cornell University Press, New York.

Bergheim, A. and Hesthagen, T. (1990). Production of juvenile Atlantic salmon, *Salmo salar* L., within different sections of a small enriched Norwegian river. *Journal of Fish Biology*, **36**, 545–62.

Berry, R. J. and Johnston, J. L. (1980). *The natural history of Shetland*. Collins, London.

Blaustein, A. R. and Wake, D. B. (1990). Declining amphibian populations: a global phenomenon? *Trends in Ecology and Evolution*, **5**, 203.

Boetius, I. and Boetius, J. (1985). Lipid and protein content in *Anguilla anguilla* during growth and starvation. *Dana*, **4**, 1–17.

Bradbury, J. W. and Vehrencamp, S. L (1976). Social organisation and foraging in Emballonurid bats. II. A model for the determination of group size. *Behavioral Ecology and Sociobiology*, **2**, 1–17.

Breen, P. A., Carson, T. A., Foster, J. B., and Stewart, E. A. (1982). Changes in subtidal community structure associated with British sea otter transplants. *Marine Ecological Programme Series*, **7**, 13–20.

Broekhuizen, S. (1989). Belasting van otters met zware metalen en PCB's. *De Levende Natuur*, **90**, 43–7.

Brzezinski, M., Jedrzejewski, W., and Jedrzejewska, B. (1993). Diet of otters (*Lutra lutra*) inhabiting small rivers in the Bialowieza National Park, eastern Poland. *Journal of Zoology, London*, **230**, 495–501.

Bunyan, P. J., Stanley, P. I., Blundell, C. A., Wardall, G. L., and Tarrant, K. A. (1975). The investigation of pesticide and wildlife incidents. *Reports of the Pest Infestation Control Laboratory 1971–1973*. Ministry of Agriculture, Guildford.

Buxton, A. (1946). *Fisherman naturalist*. Collins, London.

Carr, G. M. and Macdonald, D. W. (1986). The sociability of solitary foragers: a model based on resource dispersion. *Animal Behaviour*, **34**, 1540–9.

Carr, R. A., Jones, M. M., and Russ, E. R. (1974). Anomalous mercury in near-bottom water of a mid-Atlantic rift valley. *Nature, London*, **251**, 249.

Carss, D. N., Kruuk, H., and Conroy, J. W. H. (1990). Predation on adult Atlantic Salmon, *Salmo salar*, by otters *Lutra lutra* within the River Dee system, Aberdeenshire, Scotland. *Journal of Fish Biology*, **37**, 935–44.

Caughley, G. (1977). *Analysis of vertebrate populations*. Wiley, London.

Chanin, P. R. F. (1981). The diet of the otter and its relations with the feral mink in two areas of south-west England. *Acta Theriologica*, **26**, 83–95.

Chanin, P. R. F. (1985). *The natural history of otters*. Croom Helm, London.

Chanin, P. R. F. (1991). Otter *Lutra lutra*. In *The Handbook of British Mammals* (3rd edn) (ed. G. B. Corbet and S. Harris) pp. 423–31. Blackwell Scientific Publications, Oxford.

Chanin, P. R. F. and Jefferies, D. J. (1978). The decline of the otter *Lutra lutra* L. in Britain: an analysis of hunting records and discussion of causes. *Biological Journal of the Linnaean Society*, **10**, 305–28.

Chapman, P. J. and Chapman, L. L. (1982). *Otter survey of Ireland, 1980–81*. Vincent Wildlife Trust, London.

Cheeseman, C. L., Wilesmith, J. W., Ryan, J., and Mallinson, P. J. (1987). Badger population dynamics in a high-density area. *Symposia of the Zoological Society of London*, **58**, 279–94.

Chéhebar, C. E. (1985). A survey of the southern river otter *Lutra provocax* Thomas in Nahuel Huapi National Park, Argentina. *Biological Conservation*, **32**, 299–307.

Clutton-Brock, T. H. (1982). The function of antlers. *Behaviour*, **79**, 108–25.

Clutton-Brock, T. H. (1991) *The evolution of parental care*. Princeton University Press.

Clutton-Brock, T. H. and Harvey, P. H. (1977). Primate ecology and social organization. *Journal of Zoology, London*, **183**, 1–39.

Clutton-Brock, T. H., Guinness, F. E., and Albon, S. D (1982). *Red deer: behaviour and ecology of two sexes*. Edinburgh University Press.

Conroy, J. W. H. and French, D. D. (1987). The use of spraints to monitor populations of otters (*Lutra lutra* L.). *Symposia of the Zoological Society of London*, **58**, 247–62.

Conroy, J. W. H. and Jenkins, D. (1986). Ecology of otters in northern Scotland. VI. Diving times and hunting success otters at Dinnet Lochs, Aberdeenshire and in Yell Sound, Shetland. *Journal of Zoology, London*, **209**, 341–46.

Conroy, J. W. H., Watt, J., Webb, J. B., and Jones, A. (1993). *A guide to the identification of prey remains in otter spraint*. Occasional Publication, **16**, The Mammal Society, London.

Corbet, G. B. and Harris, S. (1991). *The handbook of British mammals*. Blackwell Scientific, Oxford.

Corbett, L. K. (1979). Feeding ecology and social organisation of wild cats (*Felis sylvestris*) and domestic cats (*Felis catus*) in Scotland. PhD Thesis, University of Aberdeen.

Costa, D. P. (1982). Energy, nitrogen, and electrolyte flux and sea-water drinking in the sea otter *Enhydra lutris*. *Physiological Zoology*, **55**, 35–44.

Costa, D. P. and Kooyman, G. L. (1982). Oxygen consumption, thermoregulation, and the effect of fur oiling and washing on the sea otter *Enhydra lutris*. *Canadian Journal of Zoology*, **60**, 2761–7.

Crawford, A. K., Jones, A., Evans, D., and McNulty, J. (1979). *Otter survey of Wales 1977–1978*. Society for the Promotion of Nature Conservation, Nettleham, Lincoln.

Creel, S. R. and Creel, N. M. (1991). Energetics, reproductive suppression and obligate communal breeding in carnivores. *Behavioral Ecology and Sociobiology*, **28**, 263–70.

Danilov, P. I. and Tumanov, I. L. (1975). The reproductive cycles of some Mustelidae species. *Byulleten Moskovskogo Obshchestva Ispytatelei* (*Oto. Biol.*), **80**, 35–47.

Davies, J. M. (1981). Survey of trace elements in fish and shellfish landed at Scottish ports, 1975–76. *Scottish Fisheries Research Reports*, **19**, 1–28.

Davies, N. (1992). *The dunnet*. Oxford University Press.

Dawson, T. J. and Fanning, F. D. (1981). Thermal and energetic problems of semi-aquatic mammals: a study of the Australian water rat, including comparisons with the platypus. *Physiological Zoology*, **54**, 285–6.

Delibes, M. and Adrian, I. (1987). Effects of crayfish introduction on otter *Lutra lutra* food in the Do:ana National Park, S. W. Spain. *Biological Conservation*, **42**, 153–9.

Derenne, P. and Mougin, J. L. (1976). Données écologiques sur les mammifères introduits de l'Ile aux Cochons Archipel Crozet (46°06'S, 50°14'E). *Mammalia*, **40**, 23–53.

Desai, J. H. (1974). Observations on the breeding habits of the Indian smooth otter *Lutrogale perspicillata* in captivity. *International Zoo Yearbook*, **14**, 123–4.

Dunbar, I. F. (1977). Olfactory preferences in dogs: the response of male and female beagles to conspecific odors. *Behavioral Biology*, **20**, 471–81.

Dunbar, R. I. M. (1988). *Primate social systems*. Croom Helm, London.

Dunstone, N. (1993). *The mink*. Poyser, London.

Duplaix, N. (1980). Observations on the ecology and behavior of the giant river otter *Pteronura brasiliensis* in Surinam. *Revue d'écologie: La terre et la vie*, **34**, 495–620.

Duplaix-Hall, N. (1975). River otters in captivity. In *Breeding endangered species in captivity* (ed. R. D. Martin), pp. 315–27, Academic, London.

Durbin, L. (1993). *Food and habitat utilization of otters (Lutra lutra L.) in a riparian habitat—the River Don in north-east Scotland*. PhD thesis, University of Aberdeen.

Egglishaw, H. W. (1970). Production of salmon and trout in a stream in Scotland. *Journal of Fish Biology*, **2**, 117–36.

Elliott, J. M. (1984). Growth, size, biomass and production of young migratory trout *Salmo trutta* in a Lake District stream, 1966–83. *Journal of Animal Ecology*, **53**, 979–94.

Elmhirst, R. (1938). Food of the otter in the marine littoral zone. *Scottish Naturalist*, **1938**, 99–102.

Erlinge, S. (1967). Home range of the otter *Lutra lutra* in Southern Sweden. *Oikos*, **18**, 186–209.

Erlinge, S. (1968). Territoriality of the otter *Lutra lutra* L. *Oikos*, **19**, 81–98.

Erlinge, S. (1972). The situation of the otter population in southern Sweden. *Viltrevy*, **8**, 379–97.

Estes, J. A. (1980). *Enhydra lutris. Mammalian Species, American Society of Mammalogists*, **133**.

Estes, J. A. (1989). Adaptations for aquatic living by carnivores. In *Carnivore behaviour, ecology and evolution* (ed. J. L. Gittleman), pp. 242–82, Cornell University Press, Ithaca.

Estes, J. A. and Palmisano, J. F. (1974). Sea otters: their role in structuring nearshore communities. *Science*, **185**, 1058–60.

Estes, J. A. and Van Blaricom, G. R. (1985). Sea-otters and shell-fisheries. In *Marine mammals and fisheries* (ed. J. R. Beddington, R. J. H. Beverton, and D. M. Lavigne), pp. 187–235, George Allen & Unwin, London.

Estes, J. A., Jameson, R. J., and Johnson, A. M. (1981). Food selection and some foraging tactics of sea otters. In *Proceedings of the Worldwide Furbearers Conference* (ed. J. A. Chapman and D. Pursley), pp.606–41. Worldwide Furbearers Conference, Frostburg, Maryland.

Estes, J. A., Jameson, R. J., and Rhode, E. B. (1982). Activity and prey selection in the sea otter: influence of population status on community structure. *American Naturalist*, **120**, 242–58.

Estes, J. A., Rathbun, G. B., and Van Blaricom, G. R. (1993). Paradigms for managing carnivores: the case of the sea otter. *Symposia of the Zoological Society of London*, **65**, 307–20

Estes, R. D. (1967). Predators and scavengers. *Natural History*, **76**, 20–29.

Estes, R. D. (1969). Territorial behaviour of the wildebeest (*Connochaetes taurinus* Burchell 1823). *Zeitschrift für Tierpsychologie*, **26**, 284–370.

Estes, R. D. and Goddard, J. (1967). Prey selection and hunting behavior of the African wild dog. *Journal of Wildlife Management*, **31**, 52–70.

Everett, M. J. (1990). Reedbeds—a scarce habitat. *Royal Society for the Protection of Birds Conservation Review*, **3**, 14–19.

Ewer, R. (1973). *The carnivores*. Weidenfeld & Nicolson, London.

Ewins, P. J. (1986). The ecology of black guillemots *Cepphus grylle* in Shetland. DPhil thesis, University of Oxford.

Fagen, R. (1981). *Animal play behavior*. Oxford University Press, New York.

Feltham, M. J. and Marquiss, M. (1989). The use of first vertebrae in separating and estimating the size of trout (*Salmo trutta*) in bone remains. *Journal of Zoology, London*, **219**, 13–22.

Foster-Turley, P., Macdonald, S. M., and Mason, C. F. (1990). *Otters, an action plan for conservation*. International Union for the Conservation of Nature, Gland.

Fraser, N. H. C., Metcalfe, N. B., and Thorpe, J. E. (1993). Temperature-dependent switch between diurnal and nocturnal foraging in salmon. *Proceedings of the Royal Society of London, B,* **252**, 135–9.

Garshelis, D. L., Johnson, A. M., and Garshelis, J. A. (1984). Social organization of sea otters in Prince William Sound, Alaska. *Canadian Journal of Zoology,* **62**, 2648–58.

Gorman, M. L. and Mills, M. G. L. (1984). Scent marking strategies in hyaenas (Mammalia). *Journal of Zoology, London,* **202**, 535–47.

Gorman, M. L. and Trowbridge, B. J. (1989). The role of odor in the social lives of carnivores. In *Carnivore behaviour, ecology, and evolution* (ed. J. G. Gittleman), pp.57–88, Cornell University Press, Ithaca.

Gosling, L. M. (1982) A reassessment of the function of scent marking in territories. *Zeitschrift für Tierpsychologie,* **60**, 89–118.

Gosling, L. M. (1990). Scentmarking by resource holders: alternative mechanisms for advertising the costs of competition. In *Chemical signals in vertebrates, 5,* (ed. D. Macdonald, D. Müller-Schwartze, and S. Natynczuk). Oxford University Press.

Grant, T. R. (1983). Body temperatures of free-ranging platypuses, *Ornithorhynchus anatinus* (Montremata) with observations on their use of burrows. *Australian Journal of Zoology,* **31**, 117–22.

Grant, T. R. and Dawson, T. J. (1978). Temperature regulation in the platypus, *Ornithorhynchus anatinus. Physiological Zoology,* **51**, 1-6, and 315–32.

Green, J. (1977). Sensory perception in hunting otters, *Lutra lutra* L. Otters, *Journal of the Otter Trust,* **1977**, 13–16.

Green, J. and Green, R. (1980). *Otter survey of Scotland, 1977-1979.* Vincent Wildlife Trust, London.

Green, J. and Green, R. (1987). *Otter survey of Scotland, 1984-1985.* Vincent Wildlife Trust, London.

Green, J., Green, R., and Jefferies, D. J. (1984). A radio-tracking survey of otters *Lutra lutra* on a Perthshire river system. *Lutra,* **27**, 85–145.

Hall, K. R. L. and Schaller, G. B. (1964). Tool-using behavior of the California sea-otter. *Journal of Mammalogy,* **45**, 287–98.

Harris, C. J. (1968). *Otters: a study of recent Lutrinae.* Weidenfeld & Nicholson, London.

Harris, S. and Smith, G. C. (1987). Demography of two urban fox (*Vulpes vulpes*) populations. *Journal of Applied Ecology,* **24**, 75–86.

Harvie-Brown, J. A. and Buckley, T. E. (1892). *A vertebrate fauna of Argyll and the Inner Hebrides.* Douglas, Edinburgh.

Heggberget, T. M. (1984). Age determination in the European otter *Lutra lutra. Zeitschrift für Säugetierkunde,* **49**, 299–305.

Heggenes, J., Krog, O. M. W., Lindas, O. R., Dokk, J. G., and Bremner, T. (1993). Homeostatic behavioural responses in a changing environment: brown trout (*Salmo trutta*) become nocturnal in winter. *Journal of Applied Ecology,* **62**, 295–308.

Herfst, M. (1984). Habitat and food of the otter *Lutra lutra* in Shetland. *Lutra,* **27**, 57–70.

Hewson, R. (1969). Couch building by otters *Lutra lutra. Journal of Zoology, London,* **195**, 554–6.

Hillegaart, V., Sandegren, F. and Ostman, J. (1981). Area utilization and marking behaviour among two captive otter (*Lutra lutra* L.) pairs. *Otters,* **1**, 64–74.

Hofer, H. and East, M. L. (1993). The commuting system of Serengeti spotted hyaenas: how a predator copes with migratory prey. *Animal Behaviour,* **46**, 547–57, 559–74, 575–89.

Irving, L. and Hart, S. (1957). The metabolism and insulation of seals as bare-skinned mammals in cold water. *Canadian Journal of Zoology,* **35**, 497–511.

Iverson, J. A. and Krog, J. (1973). Heat production and body surface area in seals and sea otters. *Norwegian Journal of Zoology,* **21**, 51–4.

Ivlev, V. S. (1966). The biological productivity of waters. *Journal of the Fisheries Research Board Canada,* **23**, 1727–59.

Jarman, P. J. (1974). The social organization of antelope in relation to their ecology. *Behaviour,* **48**, 215–66.

Jarman, P. J. and Kruuk, H. (1993) Phylogeny and spatial organization in mammals. p. 146, *Abstracts of the Sixth International Theriological Conference, University of New South Wales, Sydney.*

Jarman, P. J. and Sinclair, A. R. E. (1979). Feeding strategy and the pattern of resource-partitioning in ungulates. In *Serengeti: dynamics of an ecosystem* (ed. A. R. E. Sinclair and M. Norton-Griffiths), pp. 130–63, University of Chicago Press.

Jefferies, D. J. (1969). Causes of badger mortality in eastern counties of England. *Journal of Zoology, London,* **157**, pp. 429–36.

Jefferies, D. J. (1989). The changing otter population of Britain 1700–1989. *Biological Journal of the Linnaean Society,* **38**, 61–9.

Jefferies, D. J., French, M. C., and Stebbings, R. E. (1974). *Pollution and mammals.* Institute of Terrestrial Ecology, Huntingdon.

Jefferies, D., Green, J., and Green, R. (1984). *Commercial fish and crustacean traps: a serious cause of otter (Lutra lutra) mortality in Britain and Europe.* Vincent Wildlife Trust, London.

Jenkins, D. (1980). Ecology of otters in northern Scotland I. Otter (*Lutra lutra*) breeding and dispersion in mid-Deeside, Aberdeenshire in 1974–1979. *Journal of Animal Ecology,* **49**, 713–35.

Jenkins, D. (1981). Ecology of otters in northern Scotland. IV. A model scheme for *Lutra lutra* L. in a freshwater system in Aberdeenshire. *Biological Conservation,* **20**, 123–32.

Jenkins, D. and Burrows, G. O. (1980). Ecology of otters in northern Scotland, III. The use of faeces as indicators of otter (*Lutra lutra*) density and distribution. *Journal of Animal Ecology,* **49**, 755–74.

Jenkins, D. and Harper, R. J. (1980). Ecology of otters in northern Scotland II. Analysis of otter (*Lutra lutra*) and mink (*Mustela vison*) faeces from Deeside, N. E. Scotland, in 1977–78. *Journal of Animal Ecology,* **49**, 737–54.

Jenkins, D. and Shearer, W. M. (ed.) (1986). *The status of the Atlantic salmon in Scotland.* Institute of Terrestrial Ecology, Huntingdon.

Jenkins, D., Watson, A., and Miller, G. R. (1964). Predation and red grouse populations. *Journal of Applied Ecology*, **36**, 183–95.

Jensen, S., Kihlstrom, J. E., Olsson, M., Lundberg, C., and Ordberg, J. (1977). Effects of PCB and DDT on mink *(Mustela vison)* during the reproductive season. *Ambio*, **6**, 239.

Jolly, G. M. (1969). The treatment of errors in aerial counts of wildlife populations. *East African Agricultural and Forestry Journal*, **34**, 50–5.

Kain, J. M. (1979). A view of the genus Laminaria. *Oceanography and Marine Biology, an Annual Review*, **17**, 101–61.

Kenyon, K. W. (1959). The sea otter. *Annual Reports of the Smithsonian Institute*, **1959**, 399–407.

Kenyon, K. W. (1969). The sea otter in the Eastern Pacific Ocean. *North American Fauna, United States Department of the Interior*, **68**, 1–352.

Kitching, J. A. (1941). Studies in sub-littoral ecology. III. Laminaria forest on the west coast of Scotland; a study of zonation in relation to wave action and illumination. *Biological Bulletin, Marine Biological Laboratory, Woods Hole, Massachusetts*, **80**, 324–37.

Koeman, J. H., van de Ven, W. S. M., de Goeij, J. J. M., Tijoe, P. S., and van Haaften, J. L. (1975). Mercury and selenium in marine mammals and birds. *Science of the Total Environment*, **3**, 279–87.

Koop, J. H. and Gibson, R. N. (1991). Distribution and movements of intertidal butterfish *Pholis gunnellus*. *Journal of the Marine Biological Association of the United Kingdom*, **71**, 127–36.

Kotrschal, K., Whitear, M., and Adam, H. (1984). Morphology and histology of the anterior dorsal fin of *Gaidropsarus mediterraneus* (Pisces: Teleostei), a specialized sensory organ. *Zoomorphology*, **104**, 365–72.

Kramer, D. L. (1988). The behavioral ecology of air breathing by aquatic animals. *Canadian Journal of Zoology*, **66**, 89–94.

Krebs, J. R. (1978). Optimal foraging decision rules for predators. In *Behavioural ecology: an evolutionary approach* (ed. J. R. Krebs and N. B. Davies), pp. 23–63. Sinauer, London.

Kruuk, H. (1972). *The spotted hyena, a study in predation and social behavior.* University of Chicago Press.

Kruuk, H. (1975). Functional aspects of social hunting by carnivores. In *Function and evolution in behaviour* (ed. G. Baerends, C. Beer, and A. Manning), pp. 119–41 Clarendon, Oxford.

Kruuk, H. (1978). Foraging and spatial organisation of the European badger, *Meles meles* L. *Behavioral Ecology and Sociobiology*, **4**, 75–89.

Kruuk, H. (1986). Interactions between Felidae and their prey species: a review. In *Cats of the world: biology, conservation and management* (ed. S. D. Miller and D. D. Everell), pp. 353–73, National Wildlife Federation, Washington, D C.

Kruuk, H. (1989a). *The social badger.* Oxford University Press.

Kruuk, H. (1989b). Factors limiting populations of otters *Lutra lutra* in Shetland. *Abstracts of the Fifth International Theriological Congress, University of Rome*, p. 707.

Kruuk, H. (1992). Scent marking by otters (*Lutra lutra*): signalling the use of resources. *Behavioral Ecology*, **3**, 133–40.

Kruuk, H. (1993). The diving behaviour of the platypus (*Ornithorhynchus anatinus*) in waters with different trophic status. *Journal of Applied Ecology*, **30**, 592–8.

Kruuk, H. and Balharry, D. (1990). Effects of seawater on thermal insulation of the otter, *Lutra lutra* L. *Journal of Zoology, London*, **220**, 405–15.

Kruuk, H. and Conroy, J. W. H. (1991). Mortality of otters *Lutra lutra* in Shetland. *Journal of Applied Ecology*, **28**, 83–94.

Kruuk, H. and Conroy, J. W. H. (1987). Surveying otter *Lutra lutra* populations: a discussion of problems with spraints. *Biological Conservation*, **41**, 179–83.

Kruuk, H. and Goudswaard, P. C. (1990). Effects of changes in fish populations in Lake Victoria on the food of otters (*Lutra maculicollis* Schinz and *Aonyx capensis* Lichtenstein). *African Journal of Ecology*, **28**, 332–9.

Kruuk, H. and Hewson, R. (1978). Spacing and foraging of otters (*Lutra lutra*) in a marine habitat. *Journal of Zoology, London*, **185**, 205–12.

Kruuk, H. and Macdonald, D. (1985). Group territories of carnivores: empires and enclaves. In *Behavioural ecology: ecological consequences of adaptive behaviour* (ed. R. M. Sibley and R. H. Smith), pp. 521–36, Blackwell, Oxford.

Kruuk, H. and Mills, M. G. L. (1983). Notes on food and foraging of the honey badger *Melivora capensis* in the Kalahari Gemsbok National Park. *Koedoe*, **26**, 153–7.

Kruuk, H. and Moorhouse, A. (1990). Seasonal and spatial differences in food selection by otters *Lutra lutra* in Shetland. *Journal of Zoology, London*, **221**, 621–37.

Kruuk, H. and Moorhouse, A. (1991). The spatial organization of otters (*Lutra lutra* L.) in Shetland. *Journal of Zoology, London*, **224**, 41–57.

Kruuk. H. and Parish, T. (1981). Feeding specialization by the European badger *Meles meles* in Scotland. *Journal of Animal Ecology*, **50**, 773–88.

Kruuk, H. and Parish, T. (1987). Changes in the size of groups and ranges of the European badger (*Meles meles*) in an area in Scotland. *Journal of Animal Ecology*, **56**, 351–64.

Kruuk, H., Gorman, M., and Parish T. (1980). The use of 65-Zn for estimating populations of carnivores. *Oikos*, **34**, 206–8.

Kruuk, H., Glimmerveen, U., and Ouwerkerk, E. (1985). The effects of depth on otter foraging in the sea. *Institute of Terrestrial Ecology, Annual Report*, **1984**, 112–5.

Kruuk, H., Conroy, J. W. H., Glimmerveen, U., and Ouwerkerk, E. (1986). The use of spraints to survey populations of otters (*Lutra lutra*). *Biological Conservation* **35**, 187–94.

Kruuk, H., Conroy, J. W. H., and Moorhouse, A. (1987). Seasonal reproduction, mortality and food of otters *Lutra lutra* L. in Shetland. *Symposia of the Zoological Society of London*, **58**, 263–78.

Kruuk, H., Nolet, B., and French, D. (1988). Fluctuations in numbers and activity of inshore demersal fishes in Shetland. *Journal of the Marine Biological Association of the United Kingdom*, **68**, 601–17.

Kruuk, H., Moorhouse, A., Conroy, J. W. H., Durbin, L. and Frears, S. (1989). An estimate of numbers and habitat preference of otters *Lutra lutra* in Shetland, U. K. *Biological Conservation*, **49**, 241–54.

Kruuk, H., Wansink, D., and Moorhouse, A. (1990). Feeding patches and diving success of otters (*Lutra lutra* L.) in Shetland. *Oikos* **57**, 68–72.

Kruuk, H., Conroy, J. W. H., and Moorhouse, A. (1991). Recruitment to a population of otters (*Lutra lutra*) in Shetland, in relation to fish abundance. *Journal of Applied Ecology*, **28**, 95–101.

Kruuk, H., Carss, D. N., Conroy, J. W. H. and Durbin, L. (1993*a*). Otter (*Lutra lutra* L.) numbers and fish productivity in rivers in N.E. Scotland. *Symposia of the Zoological Society of London*, **65**, 171–91.

Kruuk, H., Kanchanasaka, B., O'Sullivan, S., and Wanghongsa, S. (1993*b*). Identification of tracks and other signs of three species of otter, *Lutra lutra, L. perspicillata* and *Aonyx cinerea*. *Natural History Bulletin of the Siam Society*, **41**, 23–30

Kruuk, H., Kanchanasaka, B., O'Sullivan, S., and Wanghongsa, S. (1994*a*). Niche separation in three sympatric otters *Lutra perspecillata, L. lutra* and *Aonyx cinerea*. *Biological Conservation*, **69**, 115–20.

Kruuk, H., Balharry, E., and Taylor, P.T. (1994*b*). The effect of water temperature on oxygen consumption of the Eurasian otter *Lutra lutra*. *Physiological Zoology*, **67** (5), in press.

Kyne, M. J., Small, C. M., and Fairley, J. S. (1989). The food of otters *Lutra lutra* in the Irish Midlands and a comparison with that of mink *Mustela vison* in the same region. *Proceedings of the Royal Irish Academy (B)*, **89**, 33–46.

Lack, D. (1954). *The natural regulation of animal numbers*. Oxford University Press.

Lampe, R. P. (1978). Aspects of the predation strategy of the North American badger *Taxidea taxus*. PhD thesis, University of Minneapolis.

Larsen, D. N. (1983). Habitats, movements and foods of river otters in coastal south-eastern Alaska. MSc thesis, University of Alaska, Fairbanks.

Latour, P. B. (1988). The individual within the group territorial system of the European badger (*Meles meles* L.). PhD thesis, University of Aberdeen.

Le Cren, E. D. (1951). The length–weight relationship and seasonal cycle in gonad weight and condition in the perch *Perca fluviatilis*. *Journal of Animal Ecology*, **20**, 201–19.

Lejeune, A. (1989). Les loutres, *Lutra maculicollis* Lichtenstein, et la Peche artisanale au Rwanda. *Revue Zoologique Africaine*, **103,** 215–23.

Lenton, E. J., Chanin, P. R. F., and Jefferies, D. J. (1980). *Otter survey of England, 1977–79.* Nature Conservancy Council, London.

Leyhausen, P. (1956). Verhaltensstudien an Katzen. *Zeitschrift für Tierpsychologie*, **suppl. 2**, 1–120.

Libois, R. M. and Rosoux, R. (1989). Ecologie de la loutre (*Lutra lutra*) dans le Marais Poitevin, I. Etude de la consommation d'anguilles. *Vie Milieu, **39**, 191–7.

Lightfoot, A. (1981). Coastal otters in Norway. Vincent Wildlife Trust, London.

Lindquist, O. (1985). Atmospheric mercury—a review. *Tellus*, **37B**, 136–59.

Loudon, A. and Racey, P. (ed.) (1987). Reproductive energetics in mammals. *Symposia of the Zoological Society of London*, **57**, 1–371.

MacArthur, R. A. (1979). Seasonal patterns of body temperature and activity in free-ranging muskrats (*Ondatra zibethicus*). *Canadian Journal of Zoology*, **57**, 25–33.

MacArthur, R. A. (1984). Aquatic thermoregulation in the muskrat (*Ondatra zibethicus*): energy demands of swimming and diving. *Canadian Journal of Zoology*, **68**, 241–8.

MacArthur, R. A. and Dyck, A. P. (1990). Aquatic thermoregulation of captive and free-ranging beavers (*Castor canadensis*). *Canadian Journal of Zoology*, **68**, 2409–16.

Macaskill, B. (1992). *On the swirl of the tide*. Jonathan Cape, London.

McCleneghan, K. and Ames, J. A. (1976). A unique method of prey capture by the sea otter, *Enhydra lutra* Linnaeus. *Journal of Mammalogy*, **57**, 410–2.

Macdonald, D. W. (1980*a*). Patterns of scent marking with urine and faeces among carnivore communities. *Symposia of the Zoological Society of London*, **45**, 107–39.

Macdonald, D. W. (1980*b*). Social factors affecting reproduction amongst red foxes. In *The Red Fox: Symposium on Behaviour and Ecology* (ed. E. Zimen), pp. 123–75 Junk, The Hague.

Macdonald, D. W. (1981). Resource dispersion and the social organization of the red fox, *Vulpes vulpes*. In *Proceedings of the Worldwide Furbearer Conference* (ed. J. A. Chapman and D. Ursley), pp.918–49, Worldwide Furbearers Conference, Frostburg, Maryland.

Macdonald, D. W. (1983). The ecology of carnivore social behaviour. *Nature, London*, **301**, 379–84.

Macdonald, D. W. (ed.) (1984). *The encyclopedia of mammals*. Allan & Unwin, London.

Macdonald, D. W. (1985). The carnivores: order Carnivora. In *Social odours in mammals* (ed. R. E. Brown and D. W. Macdonald), pp. 619–722, Clarendon, Oxford.

Macdonald, D. W. (1987). *Running with the fox*. Unwin Hyman, London.

Macdonald, D. W. and Moehlman. P. D. (1983). Cooperation, altruism , and restraint in the reproduction of carnivores. In *Perspectives in ethology, 5* (ed. P. Bateson and P. Klopfer), pp. 433–67. Plenum, New York.

Macdonald, S. M. and Mason, C. F. (1987). Seasonal marking in an otter population. *Acta Theriologica*, **32**, 449–62.

Mason, C. F. (1989). Water pollution and otter distribution: a review. *Lutra*, **32**, 97–131.

Mason, C. F. and Macdonald, S. M. (1980). The winter diet of otters (*Lutra lutra*) on a Scottish sea loch. *Journal of Zoology, London*, **192**, 558–61.

Mason, C. F. and Macdonald, S. M. (1986). *Otters, conservation and ecology*. Cambridge University Press.

Mason, C. F. and Macdonald, S. M. (1987). The use of spraints for surveying otter *Lutra lutra* populations: an evaluation. *Biological Conservation*, **41**, 167–77.

Mason, C. F. and Macdonald, S. M. (1993). PCBs and organochlorine pesticide residues in otter (*Lutra lutra*) spraints from Welsh catchments and their significance to otter conservation strategies. *Aquatic Conservation: Marine and Freshwater Ecosystems*, **3**, 43–51.

Mason, C. F. and Madsen, A. B. (1993). Organochlorine pesticide residues and PCBs in Danish otters (*Lutra lutra*). *The Science of the Total Environment*, **73**, 73–81.

Mead, R. A. (1989). The physiology and evolution of delayed implantation in Carnivores. In *Carnivore behavior, ecology and evolution* (ed. J. L. Gittleman), pp. 437–64, Cornell University Press, Ithaca.

Mech, D. (1970). *The wolf: the ecology and behavior of an endangered species.* Natural History Press, New York.

Melquist, W. E. and Dronkert, A. E. (1987). River otter. In *Wild Furbearer Management and Conservation in North America* (ed. M. Novak, J. A. Baker, M. E. Obbard, and B. Malloch), pp. 627–41, Ministry of Natural Resources, Ontario.

Melquist, W. E. and Hornocker, M. G. (1979). Methods and techniques for studying and censusing river otter populations. *University of Idaho Forest, Wildlife and Range Experimental Station, Technical Report 8.*

Melquist, W. E. and Hornocker, M. G. (1983). Ecology of river otters in west central Idaho. *Wildlife Monographs*, **83**, 1–60.

Mills, M. G. L. (1982). Factors affecting group size and territory size of the brown hyaena, *Hyaena brunnea*, in the southern Kalahari. *Journal of Zoology, London*, **198**, 39–51.

Mills, M. G. L. (1990). *Kalahari hyaenas: the behavioural ecology of two species.* Unwin Hyman, London.

Mitchell-Jones, A. J., Jefferies, D. J., Twelves, J., Green, J., and Green, R. (1984). A practical system of tracking otters *Lutra lutra* using radio-telemetry and 65-Zn. *Lutra*, **27**, 71–4.

Montevecchi, W. A. and Piatt, J. (1984). Composition and energy content of mature inshore spawning capelin. *Comparative Biochemistry and Physiology*, **78A**, 15–20.

Moorhouse, A. (1988). *Distribution of holts and their utilisation by the European otter (Lutra lutra L.) in a marine environment.* MSc thesis, University of Aberdeen.

Moors, P. J. (1980). Sexual dimorphism in the body size of mustelids (Carnivora): the roles of food habits and breeding systems. *Oikos*, **34**, 147–58.

Moriarty, C. (1978). *Eels, a natural and unnatural history.* David & Charles, London.

Moriarty, C. (1990). European catches of elver of 1928–1988. *International Review Hydrobiology*, **75**, 701–6.

Morris, P. A. (1972). A review of mammalian age determination methods. *Mammal Review*, **2**, 69–104.

Morrison, P., Rosemann, M., and Estes, J. A. (1974). Metabolism and thermo-regulation in the sea-otter. *Physiological Zoology*, **47**, 218–29.

Mortensen, E. (1977). Population, survival, growth and production of trout *Salmo trutta* in a small Danish stream. *Oikos*, **28**, 9–15.

Murphy, K. P. and Fairley, J. S. (1985). Food and spraining places of otters on the west coast of Ireland. *Irish Naturalists Journal*, **21**, 469–508.

Muus, B. J. and Dahlstrøm, P. (1974). *Collins guide to the sea-fishes of Britain and North-Western Europe.* Collins, London.

Myhre, R. and Myrberget, S. (1975). Diet of wolverine (*Gulo gulo*) in Norway. *Journal of Mammalogy*, **56**, 752–7.

Neal, E. (1986). *The natural history of badgers*. Christopher Helm, London.

Newby, T. C. (1975). A sea otter (*Enhydra lutris*) food dive record. *Murrelet*, **56**, 7.

Newman, R. M. and Waters, T. F. (1989). Differences in brown trout (*Salmo trutta*) production among contiguous sections of an entire stream. *Canadian Journal of Fisheries and Aquatic Science*, **46**, 203–13.

Newton, I., Wyllie, I., and Asher, A. (1993) Long-term trends in organochlorine and mercury residues in some predatory birds in Britain. *Environmental Pollution*, **79**, 143–51.

Nolet, B. A. and Kruuk, H. (1989). Grooming and resting of otters *Lutra lutra* in a marine habitat. *Journal of Zoology, London*, **218**, 433–40.

Nolet, B. A. and Kruuk, H. Hunting yield and daily food intake of a lactating otter (*Lutra lutra*) in Shetland. *Journal of Zoology, London*. (In press.)

Nolet, B. A., Wansink, D. E. H., and Kruuk, H. (1993). Diving of otters (*Lutra lutra*) in a marine habitat: use of depths by a single-prey loader. *Journal of Animal Ecology*, **62**, 22–32.

Noll, J. M. (1988). *Home range, movements and natal denning of river otters (Lutra canadensis) at Kelp Bay, Barranoff Island, Alaska*. MSc thesis, University of Alaska, Fairbanks.

Norman, J. R. (1963). Metabolism and thermoregulation in the sea otter. *Physiological Zoology*, **47**, 218–29.

Norton-Griffiths, M. (1973). Counting the Serengeti migratory wildebeest using two-stage sampling. *East African Wildlife Journal*, **11**, 135–49.

O'Connor, D. J. and Nielson, S. W. (1981). Environmental survey of methylmercury levels in wild mink (*Mustela vison*) and otter (*Lutra canadensis*) from the north eastern United States and experimental pathology of methylmercurialism in the otter. In *Proceedings of the World Furbearer Conference* (ed. J. A. Chapman and D. Pursley), pp. 1728–45. World Furbearer Conference, Frostburg, Maryland.

Oftedal, O. T. and Gittleman, J. L. (1989). Patterns of energy output during reproduction in carnivores. In *Carnivore behavior, ecology, and evolution*, (ed. J. L. Gittleman), pp. 355–78. Cornell University Press, Ithaca.

Olsson, M. L., Reutergårdh, L., and Sandegren, F. (1981). Var är uttern? *Sveriges Natur*, **6**, 234–40.

Osterhaus, A. D. M. E. and Vedder, E. J. (1988). Identification of virus causing recent seal deaths. *Nature, London*, **335**, 20.

Ostfeld, R. S. (1982). Foraging strategies and prey switching in the sea otter. *Oecologia*, **53**, 170–8.

Ostfeld, R. S. (1991). Measuring diving success of otters. *Oikos*, **60**, 258–60.

Ostfeld, R. S., Ebensperger, L., Klosterman, L. L., and Castilo, J. C. (1989). Foraging, activity budget and social behaviour of the South American marine otter *Lutra felina* (Molina 1782). *National Geographic Research*, **5**, 422–38.

Packer, C. and Pusey, A. E. (1984). Infanticide in carnivores. In *Infanticide: comparative and evolutionary perspectives* (ed. G. Hausfater and S. B. Hardy), pp. 31–42. Aldine, New York.

Parish, T. and Kruuk, H. (1982). The uses of radio-location combined with other techniques in studies of badger ecology in Scotland. *Symposia of the Zoological Society of London*, **49**, 291–9.

Pechlaner, H. and Thaler, E. (1983). Beitrag zur Fortpflanzungsbiologie des europäischen Fischotters (*Lutra lutra* L.). *Zoologisches Garten N. F., Jena*, **53**, 49–58.

Piotrowski, J. K. and Coleman, D. O. (1980). Environmental hazards of heavy metals: summary evaluation of lead, cadmium and mercury. *MARC Report 20*. UNEP, London.

Pond, C. (1985). Body mass and natural diet as determinants of the number and volume of adipocytes in eutherian mammals. *Journal of Morphology*, **185**, 183–93.

Powell, R. A. (1979). Mustelid spacing patterns: variations on a theme by *Mustela*. *Zeitschrift für Tierpsychologie*, **50**, 153–65.

Procter, J. (1963). A contribution to the natural history of the spotted-necked otter (*Lutra maculicollis* Lichtenstein) in Tanganyika. East African Wildlife Journal, **1**, 93–102.

Ralls, K. and Harvey, P. H. (1985). Geographic variation in size and sexual dimorphism of North American weasels. *Biological Journal of the Linnaean Society*, **25**, 119–67.

Randolph, P. A., Randolph, J. C., Mattingly, K., and Foster, M. M. (1977). Energy costs of reproduction in the cotton rat, *Sigmodon hispidus*. *Ecology*, **58**, 31–45.

Rasa, O. A. E. (1973). Prey capture, feeding techniques, and their ontogeny in the African dwarf mongoose (*Helogale undulata refula*). *Zeitschrift für Tierspychologie*, **32**, 449–88.

Reijnders, P. J. H. (1980). Organochlorine and heavy metal residues in harbour seals from the Wadden Sea and their possible effects on reproduction. *Netherlands Journal of Sea Research*, **14**, 30–65.

Reuther, C. (1980). Der Fischotter, *Lutra lutra* L. in Niedersachsen. *Naturschutz und Landschaftsforschung, Niedersachsen*, **11**, 1–182.

Reuther, C. (1991) Otters in captivity—a review with special reference to *Lutra lutra*. *Habitat*, (*Hankensbüttel*), **6**, 269–308.

Ricker, W. E. (1975). Computation and interpretation of biological statistics of fish populations. *Bulletin of the Fisheries Research Board of Canada*, **191**, 1–382.

Riedman, M. L. and Estes, J. A. (1988). Predation on seabirds by sea otters. *Canadian Journal of Zoology*, **66**, 1396–402.

Riedman, M. L. and Estes, J. A. (1990). The sea otter (*Enhydra lutris*): behavior, ecology and natural history. *United States Department of the Interior Fish & Wildlife Service* (*Washington*) *Biological Report*, **9**, pp. 1–126.

Robinson, J. (1969). Organochlorine insecticides and bird populations in Britain. In *Chemical fall-out: current research on persistent pesticides* (ed. M. W. Miller and G. C. Berg), pp. 113–73. Thomas, Springfield, Illinois.

Rood, J. P. (1986). Ecology and social evolution in the mongooses. In *Ecological aspects of social evolution*, (ed. D. I. Rubenstein and R. W. Wrangham), pp. 131–52. Princeton University Press.

Rowe-Rowe, D. T. (1977). Prey capture and feeding behaviour of South African otters. *Lammergeyer*, **23**, 13–21.

Rowe-Rowe, D. T. (1978). The small carnivores of Natal. *Lammergeyer*, **25**, 1–48.

Samuel, M. D., Pierce, D. J., and Garton, E. O. (1985). Identifying areas of concentrated use within the home range. *Journal of Animal Ecology*, **54**, 711–19.

Sandell, M. (1989). The mating tactics and spacing patterns of solitary carnivores. In *Carnivore behavior, ecology and evolution* (ed. J. L. Gittleman), pp. 164–82, Cornell University Press, Ithaca.

Schaller, G. B. (1972). *The Serengeti lion.* University of Chicago Press.

Schmidt-Neilsen, K. (1983). *Animal physiology: adaptation and environment.* Cambridge University Press.

Scholander, P. F., Walters, V., Hock, R., and Irving, L. (1950). Body insulation in some Arctic and Tropical mammals and birds. *Biological Bulletin*, **99**, 225–36.

Siegel, S. (1956). *Non-parametric statistics for the behavioral sciences.* McGraw-Hill, New York.

Sinclair, A. R. E. (1977). *The African buffalo.* Chicago University Press.

Skinner, J. D. and Smithers, R. H. N. (1990). *The mammals of the Southern African sub-region.* University of Pretoria.

Smith, T. G. and Armstrong, F. A. J. (1978). Mercury and selenium in ringed and bearded seal tissues from Arctic Canada. *Arctic*, **31**, 76–84.

Southwood, T. R. E. (1978). *Ecological methods.* Chapman & Hall, London.

Stephens, M. N. (1957). *The otter report. Universities Federation for Animal Welfare*, Potters Bar, London.

Stephenson, A. B. (1977). Age determination and morphological variation of Ontario otters. *Canadian Journal of Zoology*, **55**, 1577–83.

Strachan, R., Birks, J. D. S., Chanin, P. R. F., and Jefferies, D. J. (1990). *Otter survey of England 1984–1986.* Nature Conservancy Council, Peterborough.

Stubbe, M. (1969). Zur Biologie und zum Schutz des Fischotters *Lutra lutra* (L.). *Archiv für Naturschutz und Landschaftsforschung*, **9**, 315–24.

Stubbe, M. (1980). Die Situation des Fischotters in der D. D. R. In *Der Fischotter in Europa—Verbreitung, Bedrohung, Erhaltung* (ed. C. Reuther and A. Festetics), pp. 179–82. Aktion Fischotterschutz, Oderhaus.

Tabor, J. E. and Wight, H. H. (1977). Population status of river otter in Western Oregon. *Journal of Wildlife Management*, **41**, 692–9.

Tarasoff, F. J. (1974). Anatomical adaptations in the river otter, sea otter and harp seal with reference to thermal regulation. In *Functional anatomy of marine mammals, Vol. 2* (ed. R. J. Harrison), pp. 111–42. Academic, London.

Taylor, P. S. and Kruuk, H. (1990). A record of an otter (*Lutra lutra*) natal den. *Journal of Zoology, London*, **222**, 689–92.

Tembrock, G. (1957). Zur Ethologie des Rotfuchses (*V. vulpes* L.) unter besonderer Berücksichtigung der Fortpflanzung. *Zoologische Garten, N. F.*, **23**, 289–532.

Tinbergen, N. (1951). *The study of instinct.* Clarendon, Oxford.

Tinbergen, N. (1960). The evolution of behaviour in gulls. *Scientific American*, **1960/12**, 118–30.

Toweill, D. E. and Tabor, J. E. (1982). River otter. In *Wild mammals of North America*, (ed. J. A. Chapman and G. A. Feldhamer), pp. 688–703. Johns Hopkins University Press, Baltimore.

Twelves, J. (1983). Otter (*Lutra lutra*) mortalities in lobster creels. *Journal of Zoology, London*, **201**, 585–8.

van der Zee, D. (1981). Prey of the Cape clawless otter (*Aonyx capensis*) in the Tsitsikama Coastal National Park, South Africa. *Journal of Zoology, London*, **194**, 467–83.

van der Zee, D. (1982). Density of Cape clawless otters *Aonyx capensis* (Schinz, 1821) in the Tsitsikama National Park. *South African Journal of Wildlife Research*, **12**, 8–13.

Verberne, G. and de Boer, J. N. (1976). Chemo-communication among domestic cats. *Zeitschrift für Tierpsychologie*, **42**, 86–109.

Verwoerd, D. J. (1987). Observations on the food and status of the Cape clawless otter (*Aonyx capensis*) at Betty's Bay, South Africa. *South African Journal of Zoology*, **22**, 33–9.

Videler, J. J. (1981). Swimming movements, body posture and propulsion in cod *Gadus morhua*. *Symposia of the Zoological Society of London*, **48**, 1–27.

Watson, H. C. (1978). *Coastal otters in Shetland*. Vincent Wildlife Trust, London.

Watt, J. P. (1991). *Prey selection by coastal otters (Lutra lutra L.)*. PhD thesis, University of Aberdeen.

Watt, J. P. (1993). Ontogeny of hunting behaviour of otters (*Lutra lutra* L.) in a marine environment. *Symposia of the Zoological Society of London*, **65**, 87–104.

Wayne, R. K., Modi, W. S., and O'Brien, S. J. (1986). Morphologic variability and asymmetry in the cheetah (*Acinonyx jubatus*), a genetically uniform species. *Evolution*, **40**, 78–85.

Wayre, P. (1974). Otters in western Malaysia. *Otter Trust Annual Report*, **1974**, 16–38.

Wayre, P. (1978). Status of otters in Malaysia, Sri Lanka and Italy. In *Otters: Proceedings of the First Working Meeting of the Otter Specialist Group, Paramaribo, Surinam, March 1977*. pp. 152–5. International Union for the Conservation of Nature and Natural Resources, Morges, Switzerland.

Wayre, P. (1979). *The private life of the otter*. Batsford, London.

Webb, J. B. (1975). Food of the otter (*Lutra lutra*) on the Somerset levels. *Journal of Zoology*, **177**, 486–91.

Weber, J.-M. (1990). Seasonal exploitation of amphibians by otters (*Lutra lutra*) in north-east Scotland. *Journal of Zoology, London*, **220**, 641–51.

Weber, J-M. and Roberts, L. (1990). A bacterial infection as a cause of abortion in the European otter, *Lutra lutra*. *Journal of Zoology, London*, **219**, 688–90.

Weir, V. and Bannister, K. E. (1973). The food of the otter in the Blakeney area. *Transactions of the Norfolk and Norwich Naturalists' Society*, **22**, 377–82.

Weir, V. and Bannister, K. E. (1978). Additional notes on the food of the otter in the Blakeney area. *Transactions of the Norfolk and Norwich Naturalists' Society*, **24**, 85–8.

Westin, L. and Aneer, G. (1987). Locomotor activity patterns of nineteen fish and five crustacean species from the Baltic Sea. *Environmental Biology of Fishes*, **20**, 49–65.

Wheeler, A. (1978). *Key to the fishes of Northern Europe.* Frederick Warne, London.

Widdowson, E. M. (1981). The role of nutrition in mammalian reproduction. In *Environmental factors in mammal reproduction* (ed. D. Gilmore and B. Cook) pp. 145–65. Macmillan, London.

Wijngaarden, A. van and van de Peppel, J. (1970). De otter, *Lutra lutra* (L.) in Nederland. *Lutra,* **12**, 1–72.

Williams, T. (1986). Thermo-regulation of the North American mink during rest and activity in the aquatic environment. *Physiological Zoology,* **59**, 293–305.

Wise, M. H., Linn, I. J., and Kennedy, C. R. (1981). A comparison of the feeding biology of mink *Mustela vison* and otter *Lutra lutra. Journal of Zoology, London,* **195**, 181–213.

Wlodek, K. (1980). Der Fischotter in der Provinz Pomorze Zachodnie (West-Pommern) in Polen. In *Der Fischotter in Europa—Verbreitung, Bedrohung, Erhaltung,* (ed. C. Reuther and A. Festetics) pp. 187–94. Aktion Fischotterschutz, Oderhaus.

Woollard, T. and Harris, S. (1990). A behavioural comparison of dispersing and non-dispersing foxes (*Vulpes vulpes*) and an evaluation of some dispersal hypotheses. *Journal of Animal Ecology,* **59**, 709–22.

Woollington, J. D. (1984). *Habitat use and movements of river otters at Kelp Bay, Baranoff Island, Alaska.* MSc thesis, University of Alaska, Fairbanks.

Wren, C. D. (1984). Distribution of metals in tissues of beaver, raccoon and otter from Ontario, Canada. *Science of the Total Environment,* **34**, 112–4.

Wren, C. D. Fischer, K. L., and Stokes, P. M. (1988). Levels of lead, cadmium and other elements in mink and otter from Ontario, Canada. *Environmental Pollution,* **52**, 193–202.

Yalden, D. W. (1993). The problems of re-introducing Carnivores. *Symposia of the Zoological Society of London,* **49**, 289–306.

Index

286